Diseases in Free-living
Wild Animals

SYMPOSIA OF THE ZOOLOGICAL SOCIETY OF LONDON

NUMBER 24

Diseases in Free-living
Wild Animals

*(The Proceedings of a Symposium held at The Zoological
Society of London on 9 and 10 May, 1968)*

Edited by

A. McDIARMID

*Agricultural Research Council
Institute for Research on Animal Diseases
Compton, Berkshire, England*

Published for

THE ZOOLOGICAL SOCIETY OF LONDON

BY

ACADEMIC PRESS

1969

ACADEMIC PRESS INC. (LONDON) LTD.

Berkeley Square House

Berkeley Square

London, W1X 6BA

U.S. Edition published by

ACADEMIC PRESS INC.

111 Fifth Avenue,

New York, New York 10003

Library of Congress Catalog Card Number: 69-16500

PRINTED IN GREAT BRITAIN BY
J. W. ARROWSMITH LTD., BRISTOL

CONTRIBUTORS

AUSTWICK, P. K. C., *Parasitology Department, Ministry of Agriculture, Central Veterinary Laboratory, Weybridge, Surrey, England* (p. 249)

BAKER, J. R., *Department of Parasitology, London School of Hygiene and Tropical Medicine, London, England* (p. 147)

BARNETT, S. F., *Department of Animal Pathology, School of Veterinary Medicine, Cambridge, England* (p. 159)

BLACKMORE, D. K., *Laboratory Animals Centre, Medical Research Council Laboratories, Carshalton, Surrey, England* (p. 199)

BROCKLESBY, D. W., *Agricultural Research Council, Institute for Research on Animal Diseases, Compton, Berkshire, England* (p. 159)

BROOKSBY, J. B., *Animal Virus Research Unit, Pirbright, Surrey, England* (p. 3)

CARBERY, J. T., *Veterinary Research Laboratory, Abbotstown, Castleknock, Co. Dublin, Eire* (p. 39)

CUERDEN, C. M., *Department of Zoology, Royal Holloway College, University of London, Englefield Green, Surrey, England* (p. 75)

DUNN, A. M., *Department of Veterinary Pathology, University of Glasgow Veterinary School, Glasgow, Scotland* (p. 221)

HERMAN, C. M., *Section of Wildlife Disease and Parasite Studies, United States Department of the Interior, Fish and Wildlife Service, Patuxent Wildlife Research Center, Laurel, Maryland, U.S.A.* (p. 177)

HUGHES, D. M., *Department of Zoology, Royal Holloway College, University of London, Englefield Green, Surrey, England* (p. 75)

JENNINGS, A. R., *Department of Animal Pathology, School of Veterinary Medicine, Cambridge, England* (p. 275)

LUMSDEN, W. H. R., *Department of Animal Health, Royal (Dick) School of Veterinary Studies, University of Edinburgh, Edinburgh, Scotland* (p. 135)

MAIR, N. S., *Public Health Laboratory, Leicester Isolation Hospital, Leicestershire, England* (p. 107)

MCCAUGHEY, W. J., *Ministry of Agriculture, Veterinary Research Laboratories, Stormont, Belfast, N. Ireland* (p. 99)

MCDIARMID, A., *Agricultural Research Council, Institute for Research on Animal Diseases, Compton, Berkshire, England* (p. 119)

OWEN, DAWN, *Laboratory Animals Centre, Medical Research Council Laboratories, Carshalton, Surrey, England* (p. 199)

RANKIN, J. DEANS, *Agricultural Research Council, Institute for Research on Animal Diseases, Compton, Berkshire, England* (p. 119)

SIMPSON, D. I. H., *Microbiological Research Establishment, Porton, Wiltshire, England* (p. 13)

TAYLOR, JOAN, *Salmonella Reference Laboratory, Central Public Health Laboratory, Colindale, London, England* (p. 53)

TWIGG, G. I., *Department of Zoology, Royal Holloway College, University of London, Englefield Green, Surrey, England* (p. 75)

VAUGHAN, HELEN E. N., *Ministry of Agriculture, Infestation Control Laboratory, Field Research Station, Worplesdon, Surrey, England* (p. 289)

VAUGHAN, J. A., *Ministry of Agriculture, Infestation Control Laboratory, Field Research Station, Worplesdon, Surrey, England* (p. 289)

VIZOSO, A. D., *Virus Research Unit, Medical Research Council Laboratories, Carshalton, Surrey, England* (p. 29)

WELLS, E. A., *Department of Animal Health, Royal (Dick) School of Veterinary Studies, University of Edinburgh, Edinburgh, Scotland* (p. 135)

PARTICIPANTS IN THE DISCUSSIONS

AVERY, SHIRLEY, *47 Howcroft Crescent, Finchley, London, England*

BAILEY, L., *Rothamsted Experimental Station, Rothamsted, Harpenden, Hertfordshire, England*

BEER, J. V., *Wildfowl Trust, Slimbridge, Gloucestershire, England*

BRAMLEY, P. S., *Bellmount, Belmanground, Windermere, Westmorland, England*

CAPEL-EDWARDS, MAUREEN, *Rowan, Upper Weybourne Lane, Farnham, Surrey, England.*

CAVILL, J. P., *Veterinary Investigation Centre, Tettenhall, Wolverhampton, Staffordshire, England*

CHAPMAN, R. H., *P.O. Box 312, Orbost, 3888, Victoria, Australia*

CHU, H. P., *Department of Animal Pathology, School of Veterinary Medicine, University of Cambridge, England*

COOPER, J. E., *Quantocks, New Wokingham Road, Crowthorne, Berkshire, England*

COX, F. E. G., *Department of Zoology, King's College, London, England*

DELL, V. T. W., *90 Cheyne Way, Farnborough, Hampshire, England*

GREEN, J. B., *Virus Unit, Department of Biological Sciences, Ewell Technical College, Ewell, Surrey, England*

HAIG, D. A., *Institute for Research on Animal Diseases, Compton, Berkshire, England*

HALL, D. E., *B.D.H. (Research) Ltd., Godalming, Surrey, England*

HOARE, C. A., *Wellcome Museum of Medical Science, London, England*

HODKINSON, MADELINE, *Department of Biology, Nuclear Sciences Block, University of Salford, Lancashire, England*

IRVIN, A. D., *Institute for Research on Animal Diseases, Compton, Berkshire, England*

JONES, M., *Council for Nature, Zoological Gardens, Regent's Park, London, England*

MARSDEN, P. D., *London School of Hygiene and Tropical Medicine, London, England*

MOLYNEUX, D. H., *Molteno Institute, Cambridge, England*

MUNRO, A. L. S., *Marine Laboratory, Aberdeen, Scotland*

MURTON, R. K., *Ministry of Agriculture, Fisheries and Food, Infestation Control Laboratory, Worplesdon, Surrey, England*

ROBERTS, R. J., *Department of Veterinary Pathology, University of Glasgow, Glasgow, Scotland*

ROTH, H. H., *Food and Agricultural Organisation of United Nations, Rome, Italy*

ROTHSCHILD, MIRIAM, *Ashton Wold, Peterborough, Northamptonshire, England*

SPILLING, C. R., *Insect Pathology Unit, Commonwealth Forestry Institute, University of Oxford, England*

TAYLOR, K. D., *Ministry of Agriculture, Fisheries and Food, Infestation Control Laboratory, Tolworth, Surrey, England*

THOMPSON, H. V., *Ministry of Agriculture, Fisheries and Food, Infestation Control Laboratory, Worplesdon, Surrey, England*

VICKERMAN, K., *Department of Zoology, University College, London, England*

YEOMAN, G. H., *Veterinary Research Station, Wheathampstead, Hertfordshire, England*

ORGANIZER AND CHAIRMEN

ORGANIZER

A. McDiarmid, *Agricultural Research Council, Institute for Research on Animal Diseases, Compton, Berkshire, England*

CHAIRMEN OF SESSIONS

E. Cotchin, *Department of Pathology, Royal Veterinary College, London, England*

P. C. C. Garnham, *Department of Parasitology, London School of Hygiene and Tropical Medicine, London University, London, England*

W. M. Henderson, *Agricultural Research Council, Institute for Research on Animal Diseases, Compton, Berkshire, England*

C. L. Oakley, *Department of Bacteriology, Leeds University, Leeds, Yorkshire, England*

K. C. Sellers, *Animal Health Trust Livestock Research Centre, Stock, Essex, England*

FOREWORD

In 1931 Charles Elton, in his classic publication from Oxford on animal populations, stressed the fact that the field of wildlife pathology was still virtually untouched; practically no medical or veterinary interest had been taken in this subject apart from a few major health problems afflicting man or his domesticated stock. Plague, rinderpest, leptospirosis and rabies were good examples but in these days investigations were limited because of the lack of personnel and suitable techniques. With the development of certain wildlife species for use as laboratory animals, interest was further stimulated and Dr. Wells working at the William Dunn School of Pathology at Oxford, in the late thirties, showed how a naturally occurring mycobacterial disease in field voles could have an important effect on the vole population. Later, the causal organism was employed as a prophylactic vaccine against tuberculosis in man.

Over the years, the view has steadily developed that wildlife diseases cannot be separated from a consideration of their effect in man and/or domesticated stock and nowadays it is widely accepted that the problem must be looked at as a whole and not merely from one particular angle. In the past, ecologists have frequently dismissed the suggestion that disease could play any part in the well recognised natural fluctuations in wild populations. Some still do, stressing the fact that territorial determination, climatic conditions, food supply and "disturbed" habitats are perhaps far more important, but in the last fifteen years or so the dramatic effect of myxoma virus on the British rabbit population has been an excellent example of how wrong this view can sometimes be. A new virus in a fully susceptible host can achieve a fantastic killing level and what has happened once, could happen again perhaps with a different population of susceptible animals, including man.

General world interest in the subject has been amply demonstrated in the last decade. In 1962 an international conference, the first of its kind, was arranged in the Catskill mountains, north of New York City; the sponsors were the Wildlife Disease Association and the American Institute of Biological Sciences. It was apparent from this very well attended conference that there was considerable variation in the attitude of different countries towards the problem of disease in wildlife; some had well designed research programmes already in action, others had done very little, their contributions being based mainly on the enthusiasm of a few individuals working part-time on the subject. Britain unfortunately fell into the latter category. Again, in 1964, Dr. Vittoz, director of the Office International des Epizootics stressed the importance of the relationship between the causal agents of disease, vectors, wildlife, domesticated animals and man: O.I.E. has obviously been deeply interested in this problem for some time, judging by the discussions at international conferences in 1954 and 1958.

Moreover, the World Health Organization, in 1959, and the Food and Agricultural Organization, in 1962, have both encouraged the publication of information in this field. In 1963 a symposium on East African wildlife again emphasized the importance of the subject so far as African game animals were concerned. Some countries such as Switzerland already possess an Institute devoted to research on wildlife diseases and numerous countries have wholetime research workers in various State institutes applying themselves to this problem. It is all the more remarkable, therefore, that here in Britain with all our traditional interest in field sports and natural history, no arrangements have yet been made to deal with this important facet of wildlife ecology; it is only when a crisis occurs, such as the widespread outbreak of foot-and-mouth disease last winter, that people begin to ask what part, if any, does wildlife such as deer play in the over-all picture.

From the foregoing, it will perhaps be appreciated that the problem of disease in wildlife is worthy of far more extensive study. What has been done so far in this country has resulted from a very limited effort on the part of a few individuals who are interested, basically, in the problem but are engaged on other wholetime occupations. The situation certainly calls for far more than this. On 9 and 10 May, 1968, through the generosity and support of the Zoological Society of London, this symposium on "Diseases in Free-Living Wild Animals" was held at the meeting rooms of the society at Regent's Park. Considerable interest was shown in the eighteen papers which, over the two days, covered a wide range of topics. The aim of the symposium was two-fold: firstly, to bring together information pertaining to some of the more important diseases of wildlife, information which is either not published or is hidden in journals not readily accessible to many people and secondly, to bring together workers in the various fields, however varied their interests, for the mutual benefit of all concerned. I believe that through this symposium useful contacts have already been established between people in the ecological, medical and veterinary fields. It is to be hoped that as a natural sequel to this meeting and because of the progressive attitude of the Forestry Commission, our largest landowner, towards wildlife in general, in addition to the valuable work now being done by independent organizations such as the British Deer Society and the Game Research Association, both of whom I am associated with in an honorary capacity, more interest will be taken in the health and well being of our wildlife populations, not only for their own benefit but in the general interest of veterinary preventive medicine and public health.

I would like to take this opportunity to thank the Zoological Society for their kindness and hospitality, particularly Dr. Vevers, Assistant Director of Science, who stimulated me in the first instance to arrange this symposium and Mrs. René-Martin who, with the library staff, had the unenviable task of eventually transcribing the discussions from tape. I am particularly indebted to Mrs. Walters, one of my experimental officers at Compton, who has given me much help in the editing of the various papers and abstracting relevant matter from the discussions and to Dr. Brocklesby, one of my colleagues who kindly assisted in the preparation of the section on protozoa.

I am sure all the participants would like me to thank, on their behalf, the catering staff who did such a wonderful job in dealing with the demands of so many people in such a short period of time.

I am very grateful to all the authors who allowed themselves to be talked into giving such interesting papers and also to all those who took part so freely in the discussions. I would like to stress here that not all the contributions to the various discussions were in fact available to us due mainly to the fact that the recording system was not made full use of by all concerned. Finally, I am especially grateful to the five chairmen who presided over the proceedings in such an efficient manner and added their own pertinent remarks about the various subjects under discussion.

December 1968

A. McDIARMID

INTRODUCTION

I am pleased to welcome you, on behalf of The Zoological Society of London, to this symposium on Diseases in Free-living Wild Animals; I am particularly glad that we have Dr. Herman from the United States and Mr. Carbery from Eire here today to help in our deliberations.

The subject we are going to discuss is immensely important. In my capacity as Secretary to this Society I am much concerned about certain of its aspects. We have done, I suppose, as much as any other institution to try to find out what we can about the diseases which afflict the creatures we keep in captivity. However, there is still much to find out about diseases in free-living wild animals. Information obtained in this field might well provide us with useful knowledge which would help in deciding how to treat the animals in our collection in the best possible way in order to prevent disease and how to deal with them when they become ill. We always want to increase longevity under captive conditions.

We have known for many years that the wild creatures which come into our collection very rarely arrive in perfect health. This, needless to say, is a matter of considerable concern to us. Our knowledge of these matters has steadily increased over the years and the information emanating from a meeting such as this will undoubtedly help.

We also know that the spectrum of disease in wild animals is of great importance both to human and veterinary medicine. There is clear cut evidence of the spread of disease from wild animals in both directions. We have had the recent and unfortunate example of serious human disease coming from the handling of green monkeys from Africa afflicted with a hitherto unrecognized virus infection. In the veterinary field, we in this Society have been very much disturbed by the outbreak of foot-and-mouth disease in this country. It is the worst outbreak we have had for many years and it has not yet abated. The possibility of this infection spreading to some of the rare ungulates in our zoological collections cannot be ignored. Were the disease to spread to wild deer, a considerable problem for the veterinary authorities would arise.

It is clear from the titles of the abstracts that many of the matters which you are going to discuss have a considerable economic significance. For example, in this country, myxomatosis, by drastically curtailing the rabbit population, has enormously helped both agriculture and forestry. I heard quite recently that the vast vampire bat population in certain parts of South America is a considerable danger because they act as a natural reservoir of the rabies virus.

Another economic problem which is going to be discussed today is the question of ulcerative dermal necrosis in salmon. This is a serious matter and is causing a great deal of concern. Naturally one would like to know what part this disease plays in the reduction of the Atlantic salmon population. It may not be the only factor but, nevertheless, it could be a very important one.

For all these reasons, because of the interaction of diseases between free-living wild animals, domestic animals and man, it is in all our interests to find out as much as we can about disease in nature. We want as much background information as we can get and it seems to me that through this symposium and, I hope, its follow-up of the new and valuable scientific contacts which will be stimulated, we should be able to exploit what must be an immensely rich mine of scientific information.

I sincerely hope, therefore, that this will be a very successful symposium and that you will be able to gain from it as much as I believe I should be able to gain were I able to attend all the sessions.

Before declaring the meeting open, I am sure that you would like me to thank, at the start, Dr. McDiarmid for all the work he has done in organizing what I trust will be a very successful meeting.

S. ZUCKERMAN
The Zoological Society of London

CONTENTS

VIRUSES

Wild Animals and the Epizootiology of Foot-and-Mouth Disease

J. B. BROOKSBY

Arboviruses and Free-living Wild Animals

D. I. H. SIMPSON

A Red Squirrel Disease

A. D. VIZOSO

Ulcerative Dermal Necrosis of Salmonids in Ireland

J. T. CARBERY

BACTERIA

Salmonella in Wild Animals

JOAN TAYLOR

Leptospirosis in British Wild Mammals

G. I. TWIGG, C. M. CUERDEN and D. M. HUGHES

Brucellosis in Wildlife

W. J. McCAUGHEY

Pseudotuberculosis in Free-living Wild Animals
N. S. MAIR

Mycobacterial Infections in Free-living Wild Animals
J. DEANS RANKIN and A. McDIARMID

PROTOZOA

Trypanosome Infections of Wild Mammals in Relation to Trypanosome Diseases of Man and his Domestic Stock
E. A. WELLS and W. H. R. LUMSDEN

Trypanosomes of Wild Mammals in the Neighbourhood of the Serengeti National Park

J. R. BAKER

Some Piroplasms of Wild Mammals

S. F. BARNETT and D. W. BROCKLESBY

Mycotic Infections

PETER K. C. AUSTWICK

NEOPLASMS

Tumours of Free-living Wild Mammals and Birds in Great Britain

A. R. JENNINGS

Some Aspects of the Epizootiology of Myxomatosis

HELEN E. N. VAUGHAN and J. A. VAUGHAN

VIRUSES

Chairman: W. M. Henderson

CHAIRMAN'S INTRODUCTION

Dr. Brooksby's paper will deal with the part played by wild animals in the epizootiology of foot-and-mouth disease. Undoubtedly the rôle of wild animals in this connexion is of great importance and will assume even greater importance in the future in those countries where the disease is still endemic. In many areas of the world the significance of the native wild populations of animals is not yet very apparent but, with progress in the control of the disease in domesticated stock, I am sure that the rôle of wildlife will become of greater significance and will have to be studied much more intensively than it has been so far in order to determine what risk, if any, may persist by the maintenance of the infection in wild game. Again, as Sir Solly has already mentioned, one of the very disturbing features about the recent epidemic in this country was anticipating the action that might have to be taken if the disease spread into wildlife and one is thinking particularly of the various species of deer. I believe that on veterinary grounds alone, one would have to attempt to follow the policy adopted in the 1920's in the U.S.A. when many thousands of deer were slaughtered in the west of the United States because of foot-and-mouth disease spreading to them. One of the most worrying aspects about all this, quite apart from the impact to the deer, is to know how, in fact, one would implement a slaughter policy successfully in the deer population in this country.

Dr. Brooksby is the director of the Animal Virus Research Institute at Pirbright where the main responsibility is research on foot-and-mouth disease. In addition to his work at the Institute in this country, Dr. Brooksby has been directing investigations principally in Africa in relation to wild game.

1

Symp. zool. Soc. Lond. (1968) No. 24, 1–11.

WILD ANIMALS AND THE EPIZOOTIOLOGY OF FOOT-AND-MOUTH DISEASE

J. B. BROOKSBY

Animal Virus Research Institute,
Pirbright, Surrey, England

SYNOPSIS

Research on the rôle of wild animals in the epizootiology of foot-and-mouth disease is hampered by our lack of knowledge of the relative susceptibility of the various species concerned and on the significance of the results which can be obtained from neutralizing antibody in sera from wild animals. It is information of this kind, rather than the mere fact of observation of lesions in animals which have been trapped or shot, that provides the basis for future observations on the precise rôle of the wild animal in the dissemination of the disease or in its maintenance between outbreaks in domestic stock. Attempts are being made to fill these gaps in our knowledge but in the meantime field surveys of the incidence of the disease based on serum samples have been undertaken and have covered a wide range of African species. In studying the inter-relationship between wild animals and domestic stock, one is sometimes helped by the ease with which strains of foot-and-mouth disease virus can be identified as belonging to particular antigenic groups. Association between outbreaks can thus sometimes be circumstantially established. Another factor of interest in the interplay between wild and domestic stock is the adaptation of virus to particular hosts. It seems highly likely, from observations that positive sera are recovered from game in areas where no disease has been found in cattle and vice versa, that adaptation of strains to game animals (or conversely to cattle) does occur but, equally, there are areas where both cattle and wild animals are equally affected.

Although it is in African countries that the most complex situation exists, the problem is not confined to those areas although it is probably less extensive elsewhere. In Britain, for example, the problem is not a serious one. Hedgehogs have been recorded as being involved in the dissemination of virus in a field outbreak in Norfolk. Various other small mammals have been shown to be susceptible, including the coypu, but there are no records of these species being involved as infected disseminators of virus. Deer appear to play only a minor rôle, as judged by evidence from Western Europe, but here again the problem may be linked to the strains of virus involved and there is no doubt that the possibility exists that almost any unusual species may from time to time become involved in the epizootiology of the disease.

INTRODUCTION

Foot-and-mouth disease virus provides an interesting model for the study of the inter-relationship between disease in wild animals and in domestic stock. The natural variability of the virus, not only in respect of the well known antigenic variation into types and subtypes but in respect of variability of virulence in different species, leads to complex situations which, if they can be unravelled successfully, provide most valuable leads to possible epizootiological developments. Recent advances in our knowledge of the disease in cattle and, in particular, studies on the carrier state have emphasized that the virus/host relationship, even in domestic stock, is one of uneasy equilibrium where viruses may persist in apparently healthy animals, undergo antigenic change and then break forth to cause a new epizootic spread. Species adaptation in domestic stock may lead to virus attacking either cattle or swine and there is also evidence of strains which show some predilection for sheep. Modification of such strains to take in a wider range of species leads to the perpetuation of the virus in domestic populations. To determine whether the same situation exists between the many species of wild animals and to elucidate the situations which may lead to virus spreading from one population, the wild, to the other, the domestic, and vice versa, is obviously a very big task but one which would be very rewarding.

SUSCEPTIBLE SPECIES

In general, information on the susceptibility of particular species depends on the observations made over the years by veterinarians and others in the field. The quality of information, therefore, may vary quite widely and some doubt must exist on the status of particular species which have either not been frequently observed or in which the number of recorded cases is very small. There is, too, some information from zoos where the disease has appeared and spread to a number of different species but variation in the ease with which the virus might infect these species and the degree of contact between the animals cast doubt on negative results. Attempts have been made also to infect wild species in the field but at least on one occasion animals brought in for this purpose were probably immune as the result of earlier infection.

SERUM SURVEYS

Antibody surveys in sera offer another means of determining the prevalence of the disease in a particular species. Shortly, a joint paper from this Institute and from the Rhodesian Veterinary Department (Condy, Herniman and Hedger, 1968) will give an account of the results on 1300 sera collected mainly in Rhodesia but also in a few cases in other areas of Africa and the results do give some indication of the species of animals probably most commonly affected. Lees May and

Condy published details of the first part of this series in 1965 and their conclusions are confirmed in the larger survey. First, it must be realized that the definition of sera as positive or negative based on an arbitrary level transferred from work in cattle may very well be misleading. Serum samples from wild species, other than buffalo, have been lower than the titre that would have been expected in domestic stock at that stage of convalescence and an important experiment would be to follow the antibody titres in game animals for known periods after infection, so that the results of observations in the field would be more meaningful. Table I indicates the African species from which positive antibody titres have been obtained in work at Pirbright over the last five years. Asterisks have been included to indicate where there is very good evidence for the observation of clinical disease in the species concerned.

TABLE I

African species from which positive antibody titres have been obtained

(Asterisks denote observation of clinical disease)

*Buffalo		Reedbuck
Bushbuck	*Impala	Topi
*Duiker	*Kudu	Tsessebe
*Eland	Roan Antelope	*Waterbuck
Gemsbuck	*Sable Antelope	*Wildebeeste
*Grysbuck		*Warthog

It is misleading perhaps to put the animals in order in relation to the percentage of positive sera which have been recovered from each on the basis of a survey in substantially one area. In all the work done, however, buffalo seem to show the highest percentage of positive sera and also seem to develop highest antibody titres. The weakness of the survey is shown by the fact that rather few sera have been obtained from eland although there is sound evidence that this species has quite frequently been infected. Impala and kudu are the other species most frequently involved, showing antibody and also clinical signs of the disease. In the future it is planned to sample more extensively in areas where outbreaks have been known to occur.

INTER-RELATION BETWEEN GAME AND CATTLE POPULATIONS

Lees May and Condy (1965) have indicated how various factors may be involved in spread between game and cattle and some of their findings are depicted in Table II and Fig. 1. In Rhodesia, the northern area around Lake Kariba has a game population which is largely static and there is little contact between game and domestic stock because of tsetse fly belts and other factors. In this area positive results have been observed in sampling game sera, especially from buffalo and impala,

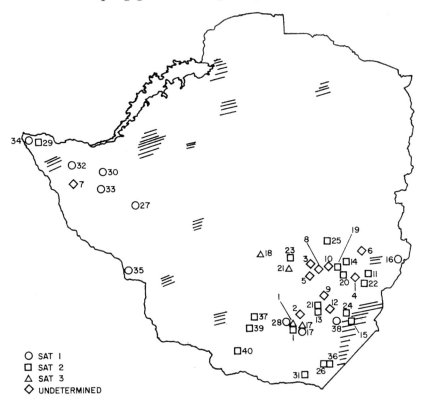

FIG. 1. Distribution of types of foot-and-mouth disease virus in cattle in Rhodesia. The shaded areas indicate areas in which game sera have been sampled. (Modified from Lees May and Condy.)

but there is no record of the disease occurring in cattle during the last 25 years. Further west, in the Wankie area, a similar situation seems to arise in that, although foot-and-mouth disease has been observed in the domestic stock from time to time, the antibodies in game animals (again especially buffalo) have been of type SAT 3 which has not appeared in cattle since regular typing of virus was started in 1950. In the south-eastern part of Rhodesia, on the large cattle ranches numbers of game graze the same pastures and use the same water points. During drought there is migration between ranches and it is in

TABLE II

Foot-and-mouth disease antibody tests
(after Lees May and Condy)

Species	No. sera tested	No. positive
Buffalo	39	18
Impala	55	4
Kudu	18	3
Sable Antelope	4	2
Roan Antelope	2	2
Wildebeeste	1	0
Eland	6	0
Warthog	23	2
Reedbuck	7	1
Tsessebe	1	0
Duiker	18	1
Steenbuck	4	0
Rhinoceros	1	0
Elephant	4	0

this area that the majority of outbreaks of foot-and-mouth disease in cattle occur. Here, too, the greatest number of positive game sera has been obtained. The types involved are generally similar in game and cattle—SAT 1 and SAT 2—although SAT 3 has also been recovered in the cattle. In other areas in Rhodesia SAT 1 has been commonest and has been found very generally in game sera and also in the associated cattle.

On several occasions in South Africa advance warning of outbreaks in domestic stock has come from the observation of the disease in game. Meeser (1962) records the recovery of the virus of type SAT 3 from impala in the Kruger National Park late in 1958. In the subsequent development of this outbreak in domestic stock many infected impala were shot on farms involved in the infection and it was suggested that the strain was particularly virulent in that species.

It has sometimes been suggested that the transfer from one species to another does result in the exaltation of virulence of the strain but work on this aspect is not quantitative and no firm conclusion can be drawn. It seems more probable that the situation is rather that the strain has adaptation to a particular species and that, when it moves from one to which it has not been preferentially adapted, it appears to have undergone exaltation.

FOOT-AND-MOUTH DISEASE IN OTHER UNGULATA

Although the greatest interest in the relation between wild and domestic stock has been in Africa, problems have arisen in temperate areas in relation to the spread from domestic stock to deer and other Cervidae. Although in an outbreak in 1926 in California a large number of deer were affected and eradication of the disease necessitated a very extensive campaign in which 70 000 deer were shot, European opinion (see, for example, Christensen, 1940) is that the disease can exist in domestic stock without much risk of infection of deer. The degree of contact between the infected animals and the deer and the question of strains of virus again may be important. It may also be that in deer the infection is very mild or inapparent, further survey work might be undertaken with advantage. There are records of the disease, based entirely on clinical evidence, in elk and reindeer but no particular epidemiological significance attaches to this, as far as is known. Many other ungulates have been considered as possibly susceptible. Elephants have been described as susceptible in one case and there are a number of reports on the susceptibility of the camel. Apart from one doubtful observation on a positive antibody titre in the serum of a hippopotamus, this species would appear to be resistant.

OTHER SUSCEPTIBLE ANIMALS

A number of small mammals of other orders than the Ungulata may be affected by the disease. McLaughlan and Henderson (1947) described the occurrence of the disease in the European hedgehog. Through the summer of 1946, hedgehogs were found in association with a series of outbreaks in Norfolk and may have been involved in spread from farm to farm. Macaulay (1963) reports the susceptibility of the East African hedgehog and has carried through the cycle of transmission, cattle–hedgehog–cattle, and extended this observation to the porcupine, which he considered to be highly susceptible and capable of excreting large amounts of virus. Recently, Capel-Edwards (1967) has demonstrated the susceptibility of the coypu and shown that the chain of infection, cattle–coypu–cattle, can be established experimentally. There is no record of coypus having been infected in this country in nature but the opportunity obviously exists for them to play a part in the dissemination of the disease. The common wild rat is marginally susceptible and it seems that any rôle it may play in the dissemination of the disease is more likely to be through mechanical carriage than as an actual infected animal. Rats ingesting virus have been shown to excrete faeces and urine containing virus for up to at least three days.

The lack of correlation between zoological classification and virus susceptibility was demonstrated very well by Snowdon (1968), who studied the susceptibility of a wide range of Australian mammals to determine their possible significance as vectors if the disease appeared in Australia. Many of the species were susceptible in the sense that virus multiplication took place and antibody was produced but lesions were seen only in a tree kangaroo, four water rats, two echidnas and possibly one red kangaroo. On the other hand, viraemia was observed in addition in grey kangaroos, wallabies, wombats, possums, potoroos and rabbits. Although in one case infection spread by contact from inoculated kangaroos to cattle, the chain of transmission, cattle–kangaroo–cattle, was not established and it is probable that susceptibility at this level implies little chance of transmission in the field.

In summary, it is suggested that the problem of foot-and-mouth disease in wild ungulates is probably similar to that in domestic stock in ranch conditions. This implies that the disease may be present in a population without any very obvious signs, the virus being harboured either in the carrier state in the pharynx of the animals concerned or occurring as mild lesions from time to time in animals with waning immunity. As new strains arise in such populations, the disease may spread to other animals of the same species. Adaptation to a particular species among those affected may at any time break down and the disease spread to related animals. It seems not improbable that in fact the two populations, wild and domestic, existing side by side, may be perpetuating the disease independently and that cross-over from one to the other may be much less common than continuing perpetuation in either of the populations concerned, except in cases where the two populations intermingle extensively.

REFERENCES

Capel-Edwards, M. (1967). Foot-and-mouth disease in *Myocastor coypus*. *J. comp. Path.* **77**, 217.

Christensen, E. H. (1940). Report on mass vaccination against foot-and-mouth disease on the Fyn Dairy district early in 1939. *Vet. Bull.* **10**, 666.

Condy, J., Herniman, K. A. J. and Hedger, R. S. (1968). Foot-and-mouth disease in wildlife in Rhodesia and other African territories: A serological survey. Submitted for publication.

Lees May, T. and Condy, J. (1965). Foot-and-mouth disease in game in Rhodesia. *Bull. Off. int. Épizoot.* **64**, 805.

Macaulay, J. W. (1963). Foot-and-mouth disease in non-domestic animals. *Bull. epizoot. Dis. Afr.* **11**, 143.

McLaughlan, J. D. and Henderson, W. M. (1947). The occurrence of foot-and-mouth disease in the hedgehog under natural conditions. *J. Hyg., Camb.* **45**, 474.

Meeser, M. J. N. (1962). Foot-and-mouth disease in game animals, with special reference to the impala (*Aepyceros melampus*). *Jl S. Af. vet. med. Ass.* **33**, 351.

Snowdon, W. A. (1968). The susceptibility of some Australian fauna to infection with foot-and-mouth disease virus. Submitted for publication.

DISCUSSION

LUMSDEN: Are the differences in virulence mentioned by Dr. Brooksby constantly related to particular antigenic types? There is evidence of such an association in the case of trypanosome antigenic types.

BROOKSBY: No, I think the question of virulence is entirely unrelated to antigenic types. You can have very virulent strains of a particular type or mildly virulent.

HAIG: At what particular age do animals show positive serum reactions and what types of reaction are seen in game?

BROOKSBY: We have no records of age of game in many of the area samplings, I regret to say. In all subsequent surveys that we are doing in cooperation with F.A.O. (Dr. Roth), we are insisting that we have full information of this kind.

TAYLOR: Is the investigation of sera for antibodies to foot-and-mouth disease virus likely to give positive results in animals which are carriers of virus but which do not suffer from the disease?

BROOKSBY: In cattle it is very unusual to get animals which are carriers without showing antibody but it has occurred. I do not think the carrier problem weakens the case for surveys of the type we are doing because they are really only revealing the top of the iceberg and there may well be something else going on down below. I think one is quite justified in carrying on with this kind of antibody survey work and hoping to supplement it later with carrier surveys. The difficulty about sampling for the carrier state is usually that you just cannot get your sample. We normally collect pharyngeal scrapings and these have to come back rather quickly to be scanned because the virus is not present in high concentration initially and is diminishing all the time in transit.

BAILEY: In view of the ease by which foot-and-mouth disease virus is transmitted mechanically and of its persistence it seems inevitable that infection will pass to wild hosts from outbreaks in domesticated animals and will persist in the wild populations. Why are these not considered important as a source of future outbreaks?

BROOKSBY: The opportunity undoubtedly exists for the virus to pass to deer but I also think that deer are probably of low susceptibility. Better opportunities have occurred in countries other than England for transmission of the disease to deer; for example, there were 181 000 outbreaks in Germany in 1951 yet there was no record at that time of any trouble in deer. The reason why it has not appeared in deer in our present outbreak is probably this low susceptibility. I think it would be useful to do a survey in the deer after things have quieted down; we considered doing a survey during the outbreak but decided we might do much more harm than good because of the possibility of the deer acting as mechanical carriers and attempts to secure specimens might entail their distribution over a wide area.

ROTH: Interpretation of antibody surveys have to be based carefully on general ecological data in respect of the sampled game populations. Age composition and physical condition, as well as specific behaviour patterns of the species concerned, have to be considered with each serum sample; therefore, F.A.O. endeavours to combine wildlife disease surveys with ecological and biological research in wildlife.

In South America, the agouti has been found to be highly susceptible to F.M.D. virus and the Pan American Foot-and-Mouth Disease Centre will publish this work in due course.

As regards elephants, there is only evidence from the Indian elephant for susceptibility to foot-and-mouth disease under unnatural conditions of close contact with diseased cattle. There is no evidence of foot-and-mouth disease in African elephants.

Symp. zool. Soc. Lond. (1968) No. 24, 13–28.

ARBOVIRUSES AND FREE-LIVING WILD ANIMALS

D. I. H. SIMPSON

Microbiological Research Establishment,
Porton, Salisbury, Wiltshire, England

SYNOPSIS

Almost all arbovirus diseases are zoonoses, the natural hosts being wild animals. Arbo-viruses are maintained in nature by multiplication in a vertebrate host with the pro-duction of a viraemia. A haematophagous arthropod feeding on the host during this viraemic stage becomes infected and during an extrinsic incubation period the virus multiplies in the tissues of the arthropod. The virus can then be passed on by bite to another susceptible host. The very wide range of vertebrate hosts include arboreal and terrestrial mammals, birds and even reptiles.

Most arboviruses, even those which cause severe and often fatal illnesses in man and domestic animals, have little visible effect upon the maintenance host. Although virus multiplication and subsequent viraemia takes place the animal does not become ill. Only in rare cases will infection with an arbovirus cause illness and death of a natural host. Examples of this are Kyasanur Forest Disease in monkeys in India and yellow fever in some South American primates. Following infection, antibody develops and the vertebrate host cannot be re-infected. Thus for virus activity to continue a sufficient number of susceptible hosts must be produced annually to maintain the vertebrate–arthropod–vertebrate cycle.

Examples of various arboviruses and their probable maintenance hosts are presented.

INTRODUCTION

An arbovirus has been defined as a virus which is maintained in nature principally through biological transmission between susceptible vertebrate hosts by haematophagous arthropods; they multipy and produce viraemia in the vertebrate, multiply in the tissues of arthropods and are passed on to new vertebrates by the bites of arthropods after a period of extrinsic incubation (W.H.O., 1967). This excludes viruses such as myxomatosis which can be transmitted mechanically on the mouth-parts of arthropods which act as an "infected pin". The definition does not, however, take into consideration the fact that some viruses classed as arboviruses may be transmitted by other mechanisms such as the oral and respiratory routes. Examples of these are the airborne trans-mission of some bat viruses (Sulkin, 1961; Constantine and Woodall, 1964), transmission of Eastern equine encephalitis virus by pecking in pheasants, transmission of Central European tick-borne encephalitis viruses in infected goats milk and transmission of Machupo virus the causative agent of Bolivian Haemmorhagic Fever in the urine of

infective rodents. Professor Chumakov (W.H.O., 1967) states that arboviruses should be defined as zoonotic viral agents which when circulating in natural foci of infection are transmitted in a more or less
regular manner by arthropods but, in certain cases, may be transmitted
in other ways than by arthropods. Smith (1968) has suggested that as
arboviruses are such a heterogenous group the most realistic definition
would be potential zoonotic viruses which cannot be otherwise classified (i.e. as enteroviruses, herpes viruses, etc).

FACTORS INFLUENCING THE SURVIVAL OF AN ARBOVIRUS

Hosts

Both arthropod and vertebrate hosts are essential for the continued
survival of an arbovirus in nature. Arthropod hosts may be divided
into 4 categories: (1) mosquitoes, (2) ticks, (3) *phlebotomus* and (4)
culicoides. *Culicoides* are thought to transmit Bluetongue virus and the
virus of African horsesickness whilst *Phlebotomus* species have yielded
nine viruses particularly Naples and Sicilian sandfly fever virus and
vesicular stomatitis virus. By far the most important arbovirus vectors
are mosquitoes and ticks. Almost all arbovirus diseases are zoonoses and
the vertebrate hosts are generally free-living wild animals. The very
wide range of such hosts includes primates, rodents, birds, bats,
ungulates, marsupials and even reptiles.

Both arthropod and vertebrate hosts may be divided into *maintenance* and incidental hosts. Maintenance hosts ensure the continued
survival and distribution of a virus but incidental hosts do not necessarily do so. Infection of incidental hosts occurs but transmission from
them is not regular enough to maintain the virus. Man is an excellent
example of an incidental vertebrate host in many arbovirus cycles
and is frequently a "dead-end" as far as virus maintenance is
concerned.

Host-virus relationship

Most arboviruses, even those which cause severe and often fatal
illness in man and domestic animals, have little visible effect upon the
maintenance host. These hosts do not show signs of infection although
they are truly "infected". Virus multiplies in their tissues and circulates
in their blood. This viraemic stage is followed by the formation of antibody to the virus rendering the animal immune to further infection with
this virus. The natural relationship between virus and host is presumably the result of evolutionary adaptation of a truly symbiotic existence.
Only in rare cases will infection with arboviruses cause recognizable

illness and death in wild animals. Examples of this are Kyasanur Forest disease in Indian monkeys and yellow fever in some species of South American primates.

Factors influencing virus transmission

Virus must circulate in the bloodstream of a vertebrate in sufficient quantity and for long enough to be able to infect a biting arthropod. Enough arthropods must also be infected to ensure the infection of other vertebrate maintenance hosts. Chamberlain, Sikes, Nelson and Sudia (1954) defined the lowest concentration of virus in the blood sufficient to infect at least 5% of a specified arthropod species as the threshold. There is a tremendous variation in the threshold value for various viruses and arthropod species. *Culex tarsalis* can be infected by amounts of St. Louis encephalitis virus barely detectable by usual laboratory means. The threshold for Western equine encephalitis virus is almost 1000 mouse LD_{50}. The duration of viraemia above the threshold level is also very important. Virus maintenance depends on the density of the arthropod population and its biting preference. In other words, if the duration of viraemia is short, a large arthropod population with a high biting frequency is required to ensure that sufficient become infected. Periods of longer viraemia require a smaller arthropod population. Mosquitoes bite fairly frequently but each feed is relatively brief. Ticks, on the other hand, often feed on the same animal over a period of several days and could overlap a viraemic stage. When an arthropod is infected, the virus must undergo a period of extrinsic incubation before the arthropod can infect another vertebrate. Even then it may feed on relatively few maintenance hosts in its lifetime. The chances of a maintenance host being bitten depends on their availability and incidental hosts are often bitten. The importance of these incidental hosts is often not concerned with virus maintenance but with arthropod maintenance. Examples are the movement of cattle in areas where Kyasanur Forest Disease and Russian Spring-Summer encephalitis viruses are prevalent. These cattle are excellent hosts for tick species allowing the tick populations to increase enormously and thus increase the risk of infection to other vertebrates.

Arthropods once infected are apparently infective for life. Transmission depends on their biting frequency, how many feed on susceptible maintenance vertebrate hosts and on the survival of the arthropod. A mosquito's life is relatively short but its high biting frequency ensures transmission to other susceptible vertebrates. Ticks can live for several years and it is well known that transtadial and transovarial transmission of tick-borne viruses can occur.

Vertebrate hosts are necessary in these cases largely to ensure survival of the tick species as well as being hosts for the virus.

As already mentioned, vertebrates once infected become immune to further infection and therefore are no longer useful in maintaining an arbovirus cycle. A constant supply of fresh, susceptible vertebrates must therefore be provided to ensure virus survival. Small mammals such as rats and voles are the hosts of many of the tick-borne complex of viruses. They breed rapidly and therefore a continuous replenishment of susceptible animals is able to keep pace with the frequency of transmission. In such small mammal populations there are foci of infection since adult species tend to live out their life in a defined area. A good example of this is demonstrated by Langat virus isolated by Smith in 1956 in a very small forest reserve in Malaya. Repeated isolations of this virus have been made from very small numbers of ticks collected in this area during the past 10 years. Transmission of such viruses outside the area of natural focus would occur if the rodent or tick population moved into fresh areas.

Mosquitoes bite frequently and rapidly exhaust the maintenance hosts in an area unless sufficient fresh hosts are constantly available. If these are not available the infection must move away from the area. Yellow fever is a good example as slowly reproducing primates are the normal maintenance vertebrates. In the tropical forests of the Congo basin and South America the areas are so vast that continuous cycles of yellow fever can persist. However, the Central American wave of yellow fever which occurred betwen 1949 and 1956 and which spread northwards from the Panama to the Mexican border exhausted all the maintenance hosts and the wave of infection petered out.

VERTEBRATE SPECIES AND THEIR ASSOCIATED ARBOVIRUSES
(Tables I, II and III)

Arboviruses which cause disease in man have naturally received the most detailed epidemiological study. These studies have elicited the probable vertebrate maintenance hosts of several viruses, particularly those causing human encephalitis, but whether these are the true maintenance hosts has still be be convincingly proved.

Birds

It is generally agreed that wild birds are the most important natural hosts of 3 important viruses occurring in North America—namely

Eastern Equine encephalitis (EEE), Western Equine encephalitis (WEE) and St. Louis encephalitis virus (SLE). EEE and WEE have been isolated in the Caribbean and Southern America whilst SLE is probably the most important cause of human encephalitis in the U.S.A. Ten Broeck (1938) first suggested birds as the reservoir host of EEE and this was supported by Kissling, Chamberlain, Sikes and Eidson (1954) who found a large number of wild birds were susceptible to experimental infection. Isolations of EEE have been made from pheasants, pigeons, cardinal, sparrow, wren, grackle, woodthrush and catbirds. The natural arthropod vector is still unknown, isolations having been made from 12 mosquito species. Oral transmission was demonstrated in pheasants by Holden (1955) without an arthropod vector. The extensive studies carried out in Kern County, California (Reeves, Hammon, Long-Shore, McClure and Beib, 1962) proved that birds were probably the basic reservoir of WEE. Of wild birds from 15 species, 17% had antibodies and experimental infection produced high levels of viraemia. Isolates have been made from prairie chicken, sparrow, red-winged blackbird and blue jays amongst others. Virus was recovered from birds 10 months after infection (Reeves, 1959) which suggests a possible means of virus maintenance through the winter period. The maintenance mosquito host is thought to be *Culex tarsalis* (Hammon, Reeves, Brookman and Izumi, 1941). SLE has caused recent epidemics of major proportions in Florida and several mid-West states. Chamberlain (1958) suggested a sylvan cycle between woodland birds and *Culex salinarius*. Infected wild birds introduce the disease into urban areas and infect *Culex pipiens* and *C. quinquefasciatus* which in turn infect the peri-domestic birds. *C. quinquefasciatus* may also infect man. Downs (1963) has reported isolation of SLE from birds in Trinidad.

Another virus of the Americas which has received considerable study is Venezuelan equine encephalitis (VEE). Although birds probably act as hosts to this virus, Chamberlain, Kissling, Stamm, Nelson and Sikes (1956), after producing only low levels of viraemia after experimental infection, did not consider that birds were the important natural reservoirs of VEE. Rodents are almost certainly the true maintenance hosts.

In the Far East and India, Japanese encephalitis virus (JE) is an important cause of encephalitis. Buescher (1956) demonstrated the importance of nestling wild birds especially black crowned night herons as the natural hosts of JE and this was confirmed by Scherer *et al.* (1959) after extensive ecological studies in the Tokyo area. Egrets were also implicated and the mosquito host was shown to be *Culex tritaeniorhynchus*.

Miles (1960), in reviewing the epidemiology of Murray Valley encephalitis, has shown that the virus exists as an enzootic cycle in birds and
mosquitoes in northern tropical Australia and in New Guinea. The virus
is carried southward to southern areas of Australia by migrating birds.

Other important viruses in which birds are the natural hosts are
West Nile (Taylor, Work, Hurlbut and Rizk, 1956) and Sindbis (Taylor,
Hurlbut, Work, Kingston and Frothingham, 1955) both of which are
widespread in Africa and have been found in the Far East and in India.

Several other viruses such as Quaranfil, Ilheus, Highlands and
Ingwavuma are known to have birds as their probable natural hosts but
extensive studies have not been carried out. Ground living birds have
been implicated in tick-borne encephalitis virus maintenance but are
not thought to play a very important rôle. Birds may, however, be
important in transporting infected ticks to new areas and Hoogstraal
and Kaiser (1961) have shown that tick species are transferred from
northern Russia to NE Africa on migrant birds. Migrating birds are
thought to be the mechanism for the wide distribution of WEE, EEE
and SLE in the Americas.

Rodents and insectivores

Small mammals such as voles and rats are the main vertebrate hosts
of the tick-borne encephalitis viruses. These viruses are perfect examples of Pavlovsky's theory of nidality (Audy, 1958) which is that a
disease tends to have a natural habitat just as a species does. The viruses
exist in distinct foci and cycles in vertebrates and ticks. Small rodents
are the most important hosts. Central European tick-borne encephalitis
has a wide variety of hosts including voles, field mice and hedgehogs.
Maintenance of the Russian Spring-Summer encephalitis complex
probably also involves squirrels, lemmings and chipmunks. Louping ill,
the only arbovirus recognized in the British Isles, may be maintained
in voles and shrews (Smith, 1962) but the tick population is probably
dependent to some extent on the rodent population for its continued
existence. Virus isolations have been made from shrews and wood mice
(Smith, Varma and McMahon, 1964). Kyasanur Forest Disease is
probably maintained in rodents and ground living birds (Work, 1958;
Webb, 1965). Other examples of this group of viruses which are dependent on rodent vertebrate hosts are Powassan (McLean and Larke, 1963),
which has been isolated from squirrels, chipmunks and groundhogs in
Canada and is known to occur in the U.S.A., and Langat (Smith, 1956),
which has been isolated from ticks and antibody found in rodents in
Malaya.

Another member of the tick-borne group of viruses is Colorado Tick Fever virus which is not related to the Russian Spring-Summer complex. Burgdorfer and Eklund (1959, 1960) isolated the virus from golden-mantled ground squirrels and Columbian ground squirrels; deermice and porcupines have also been implicated.

The Group C viruses isolated in Brazil and in Central America (Causey, Causey, Maroja and Macedo, 1961; Shope, Causey and Causey, 1961; Shope and Whitman, 1966; de Rodaniche, de Andrade and Galindo, 1964) are probably all maintained in rodents. *Oryzomys*, *Nectomys* and *Proechimys*, a sloth and some small marsupials, have all yielded strains of these viruses.

The rôle of rodents in the cycle of South American Haemorrhagic Diseases such as the Argentinian and Bolivian outbreaks caused by Junin and Machupo viruses, respectively, has been established. The sylvatic rodent *Proechimys cayannensis* appears to be infected with Machupo virus in the forest and possibly infects peridomestic species *Calomys callosus*. *Calomys* is known to develop viraemia and excretes large quantities of virus in the urine (Johnson, Mackenzie, Webb and Kuns, 1965). Junin has been isolated from field mice *Apodemus Sylvaticus*, Hesperomys and Akodon species (Greenaway *et al.*, 1959). No arthropod host has been found.

A vast number of viruses have been isolated from rodents but most of them have been incompletely studied. Daubney and Hudson (1934) suggested that *Arvicanthis* species might be the maintenance reservoir of Nairobi Sheep Disease virus and Weinbren and Mason (1957) found that *Arvicanthis* might be a possible host of Rift Valley Fever virus. Simpson (1966) showed that this species was susceptible to several African arboviruses. The rôle of the rodents in the natural cycle of EEE, WEE, SLE and several other viruses has been incompletely studied but Karstad, Spalatin and Hanson (1961) concluded that rodents, although susceptible to infection with EEE, failed to produce demonstrable viraemia and were therefore unlikely to be true maintenance hosts. It is interesting that Henderson, Karabatsos, Bourkem, Wallis and Taylor (1962) found a higher incidence of SLE antibodies in wild-caught rodents than in wild birds in a recent Florida survey. VEE and some other South American viruses are almost certainly maintained in rodent reservoirs (Causey *et al.*, 1961; Downs, Spence and Aitken, 1962).

Primates

The rôle of primates in the natural cycle of arboviruses is not clearly understood except for yellow fever virus. In Africa, monkeys are the

natural vertebrate hosts and the cycle is maintained in the forest
canopies with *Aedes africanus* as the vector. In South America monkeys
are again the vertebrate hosts but disease occurs in several species and
death results. This is probably due to the fact that yellow fever was
introduced into the Americas from Africa and complete adaptation
to the American primates has not taken place.

Many other viruses have been isolated from monkeys or antibodies
have been found in different primate species but the role of these species
as maintenance hosts is not so clear. Smith (1958) postulated that
monkeys might be a possible maintenance host of dengue in the forests
of Malaya. This was supported by Rudnick (1966) who found dengue
neutralizing antibodies in wild monkeys but the evidence is still very
inconclusive.

Leporidae

The Californian encephalitis group of viruses appear to be maintained in rabbits and hares. California encephalitis virus has been isolated from snowshoe hares (Burgdorfer, Newhouse and Thomas, 1961)
and Tahyna virus has produced viraemia when inoculated experimentally into wild rabbits and hares (Simkova, 1963). Button-willow
may also have the hare as a natural host.

Ungulates

Antibodies to vesicular stomatitis virus, which generally affects
domestic cattle, have been found in wild deer in the U.S.A. Deer have
also been found to be very susceptible to experimental infection
(Karstad and Hanson, 1957). No conclusive evidence is available to
demonstrate that deer act as maintenance hosts of this virus. *Phlebotomus* are the probable arthropod vectors.

Outbreaks of a severe haemorrhagic disease known as epizootic
haemorrhagic disease of deer have been reported in the United States
and the disease appears to be species specific.

There is some evidence that elks in Scandinavia are susceptible to
experimental infection with Central European tick-borne encephalitis
virus (Svedmyr, von Zeipel, Borg and Hansen, 1965) and Dunn (1960)
has suggested that deer in the British Isles might play a part in the
maintenance of louping-ill virus.

Bats

Several viruses which are antigenically closely related to Group B
arboviruses have been isolated from the salivary glands of various bat

species in Africa and America. Rio Bravo (Burns and Farinacci, 1956); MML (Bell and Thomas, 1964); Dakar bat (Bres and Chambon, 1963); Entebbe bat (Lumsden, Williams and Mason, 1961) and Bukalasa bat (Williams, Simpson and Shepherd, 1964) have all been isolated from insectivorous bats and all are apparently species specific. Other bat isolates have been Lagos bat virus isolated from the brain of an *Eidelon* fruit bat (Boulger and Porterfield, 1958) and Tacaribe virus from *Artibeus* bats in Trinidad (Downs, Anderson, Spence, Aitken and Greenhall, 1963). Tacaribe is related antigenically to Junin virus which causes Argentinian haemorrhagic fever. Sulkin, Allen and Sims (1963, 1964) have demonstrated the susceptibility of the Mexican free-tailed bat to JE and SLE virus and showed evidence of transplacental transmission of JE. Shepherd and Williams (1964) and D. I. H. Simpson (unpublished results) have shown that the susceptibility of various bats to experimental infection with various African arboviruses but the rôle of bats as maintenance hosts still requires considerable study. There is yet no evidence that any of the bat salivary gland viruses are arthropodborne and there is evidence of aerial transmission from bat to bat (Sulkin, 1961; Constantine and Woodall, 1964).

Marsupials, reptiles and amphibia

There is evidence that Ross River virus might be maintained in Australian kangaroos whilst Group C viruses and Apeu have been isolated from marsupials in South America. The rôle of reptiles is obscure but snakes are susceptible to EEE and WEE, whilst Chaco, Marco and Timbo viruses have been isolated from lizards (Causey, Shope and Bensabath, 1966). Karstad (1961) even produced viraemia in alligators and turtles when inoculated with EEE.

OVERWINTERING MECHANISMS

Considerable study of the mechanisms of virus survival through the cold winter months in temperate zones has been made. Whilst it is thought that tick-borne viruses can be maintained in the tick host during this period, how mosquito-vectored viruses survive is uncertain. Three possible mechanisms have been considered:

(a) persistence of infection in hibernating mosquitoes,
(b) persistence of infection in hibernating vertebrates and
(c) re-introduction of viruses in the Spring.

Support for the first mechanism was given by Blackmore and Winn (1956) who isolated WEE from naturally infected *Culex tarsalis* collected in winter months. SLE and WEE were isolated from overwintering mosquitoes by Reeves, Bellamy and Scrivani (1958) whilst Hurlbut (1950) isolated JE from mosquitoes after experimental hibernation. There is some indication that virus survival is tenuous through the winter in mosquito hosts (Bellamy, Reeves and Scrivani, 1958) and its pathogenicity may be altered.

Considerable evidence exists for the overwintering of virus in hibernating vertebrates. Gebhart, Stanton, Hill and Collett (1964) isolated WEE from naturally infected snakes following hibernation. This had previously been shown experimentally by Thomas, Eklund and Rush (1958); Thomas and Eklund (1960). La Motte (1958) and Nosek, Gresikova and Rehacek (1961) suggested bats as winter hibernating hosts of JE and tick-borne encephalitis virus, respectively. Van Tongeren (1959) produced experimental evidence of tick-borne encephalitis virus persistence in hibernating hedgehogs and this was confirmed by Kozuch, Nosek, Ernek, Lichard and Albrecht (1963) in hedgehogs and doormice. Goldfield and Sussman (1963) suggested small rodents as overwintering hosts of WEE and SLE. Reeves (1959) suggested birds as overwintering reservoirs after recovering WEE from birds 10 months after experimental infection.

Very little evidence of re-introduction of viruses into temperate regions in the Spring exists beyond the suggestion by Kissling, Stamm, Chamberlain and Sudia (1960) that viruses might be introduced by migrant birds.

The problems of overwintering have not yet been satisfactorily explained nor has the persistence of infection in tropical regions during long, dry periods when the arthropod vector populations drop markedly. It is possible the infection is maintained in a very few resting mosquitoes.

SUMMARY

Free-living wild animals are the natural vertebrate maintenance hosts of almost all arboviruses. The range of animals extends from birds to reptiles. Generally the vertebrate host shows no sign of infection with arboviruses although wild animals sometimes develop illness and occasionally die as a result of arbovirus infections but these animals are presumably not true maintenance hosts.

Many of the arboviruses known to man do not cause obvious disease and several have yet to be shown to be true arboviruses even although antigenically related.

The rôle of wild animals in the mechanisms of overwintering of arboviruses is not yet understood.

TABLE I

Distribution of 204 *arboviruses in arthropods and vertebrates*

Isolated only from arthropods	79
Isolated only from vertebrates	37
Isolated both from vertebrates and arthropods	88
	TOTAL 204*

* The total 204 is based on arboviruses registered in the Arbovirus Information Exchange Catalogue in February, 1967.

TABLE II

Distribution of arbovirus isolates by group amongst wild vertebrates and invertebrates

	Group A	Group B	Group C	Group Bunyam-wera	All other	TOTAL
Total viruses in group	20	42	13	14	115	204
Isolates from:						
Mosquitoes	19	24	8	13	60	124
Ticks	0	11	1	0	21	33
Mammals (other than Man)	7	18	9	2	35	71
Birds	4	8	0	0	7	19

TABLE III

Isolation of arboviruses from various wild vertebrate species

Vertebrate species	Number of arboviruses isolated	Number of arboviruses isolated solely from one vertebrate species
Rodents	42	19
Bats	11	7
Birds	18	5
Primates	5	1
Ungulates	4	1
Marsupials	10	2
Other Mammals	7	2
Reptiles and Amphibia	5	3
TOTAL	102	40

REFERENCES

Audy, J. R. (1958). The localization of disease with special reference to the zoonoses. *Trans. R. Soc. trop. Med. Hyg.* **52**, 308.

Bell, J. F. and Thomas, L. A. (1964). A new virus "MML" enzootic in bats (*Myotis lucifugus*) of Montana. *Am. J. trop. Med. Hyg.* **13**, 607.

Bellamy, R. E., Reeves, W. C. and Scrivani, R. P. (1958). Relationship of mosquito vectors to winter survival of encephalitis virus. II. Under experimental conditions. *Am. J. Hyg.* **67**, 90.

Blackmore, J. S. and Winn, J. F. (1956). A winter isolation of western equine encephalitis virus from hibernating *Culex tarsalis* Coquillet. *Proc. Soc. exp. Biol. Med.* **91**, 146.

Boulger, L. R. and Porterfield, J. S. (1958). Isolation of a virus from Nigerian fruit bats. *Trans. R. Soc. trop. Med. Hyg.* **52**, 421.

Bres, P. and Chambon, L. (1963). Isolement a Dakar d'une souche d'arbovirus a partis des glandes salivaires de chauve-souris. *Annls Inst. Pasteur*, Paris, **104**, 705.

Buescher, E. L. (1956). Arthropod-borne encephalitis in Japan and South-east Asia. *Am. J. pub. Hlth* **46**, 597.

Burgdorfer, W. and Eklund, C. M. (1959). Studies in the ecology of Colorado tick fever in western Montana. *Am. J. Hyg.* **69**, 127.

Burgdorfer, W. and Eklund, C. M. (1960). Colorado tick fever: I. Further ecological studies in western Montana. *J. infect. Dis.* **107**, 379.

Burgdorfer, W., Newhouse, V. F. and Thomas, L. A. (1961). Isolation of California encephalitis virus from the blood of a snowshoe hare (*Lepus americanus*) in western Montana. *Am. J. Hyg.* **73**, 344.

Burns, K. F. and Farinacci, C. J. (1956). Virus of bats antigenically related to St. Louis encephalitis. *Science, N.Y.* **123**, 227.

Causey, O. R., Causey, C. E., Maroja, O. M. and Macedo, D. G. (1961). The isolation of arthropod-borne viruses, including members of two hitherto undescribed serological groups, in the Amazon region of Brazil. *Am. J. trop. Med. Hyg.* **10**, 227.

Causey, O. R., Shope, R. E. and Bensabath, G. (1966). Marco, Timbo and Chaco, newly recognized arboviruses from lizards of Brazil. *Am. J. trop. Med. Hyg.* **15**, 1966.

Chamberlain, R. W., Sikes, R. K., Nelson, D. B. and Sudia, W. D. (1954). Studies on North American arthropod-borne encephalitis. VI. Quantitative determinations of virus-vector relationships. *Am. J. Hyg.* **60**, 278.

Chamberlain, R. W., Kissling, R. E., Stamm, D. D., Nelson, D. B. and Sikes, R. K. (1956). Venezuelan equine encephalomyelitis in wild birds. *Am. J. Hyg.* **63**, 261.

Chamberlain, R. W. (1958). Vector relationships of the arthropod-borne encephalitides in North America. *Ann. N.Y. Acad. Sci.* **70**, 312.

Constantine, D. C. and Woodall, D. F. (1964). Latent infection of Rio Bravo virus in salivary glands of bats. *Publ. Hlth Rep.*, Wash. **79**, 1033.

Daubney, R. and Hudson, J. R. (1934). Nairobi sheep disease: natural and experimental transmission by ticks other than *Rhipicephalus appendiculatus*. *Parasitology* **26**, 496.

de Rodaniche, E., de Andrade, A. P. and Galindo, P. (1964). Isolation of two antigenically distinct arthropod-borne viruses of group C in Panama. *Am. J. trop. Med. Hyg.* **13**, 839.

Downs, W. G., Spence, L. and Aitken, T. H. G. (1962). Studies on the virus of Venezuelan equine encephalomyelitis in Trinidad, W.I. III. Re-isolation of virus. *Am. J. trop. Med. Hyg.* **11**, 841.

Downs, W. G. (1963). Birds in relation to arthropod-borne viruses in Trinidad. *Int. orn. Congr.* **13**: 581–590.

Downs, W. G., Anderson, C. R., Spence, L., Aitken, T. H. G. and Greenhall, A. H. (1963). Tacaribe virus, a new agent isolated from *Artibeus* bats and mosquitoes in Trinidad, West Indies. *Am. J. trop. Med. Hyg.* **12**, 640.

Dunn, A. M. (1960). Louping-ill: The red deer (*Cervus elaphus*) as an alternative host of the virus in Scotland. *Br. vet. J.* **116**, 284.

Gebhart, L. P., Stanton, G. J., Hill, D. W. and Collett, G. C. (1964). Natural overwintering hosts of the virus of Western equine encephalitis. *New Engl. J. Med.* **271**, 172.

Goldfield, M. and Sussman, O. (1963). Verbal presentation of data obtained in epidemiological work in New Jersey with Eastern and Western encephalomyelitis. Annual fall meeting of the New Jersey Public Health Assn. Oct. 16 at Newark, N.J.

Greenaway, D. J., Rugiero, H. R., Parodi, A. S., Frigerio, M., Rivero, E., de la Barrera, J. M., Garzon, F., Boxaca, M., Mettler, N., de Guerrero, L. B. and Nota, N. (1959). Epidemic Haemorrhagic Fever in Argentina. *Publ. Hlth Rep.*, Wash. **74**, 1011.

Hammon, W. McD., Reeves, W. C., Brookman, B. and Izumi, E. M. (1941). Isolation of the virus of western equine and St. Louis encephalitis from *Culex tarsalis* mosquitoes. *Science, N.Y.* **94**, 328.

Henderson, J. R., Karabatsos, N., Bourkem, A. T. C., Wallis, R. C. and Taylor, R. M. (1962). A survey for arthropod-borne viruses in south-central Florida. *Am. J. trop. Med. Hyg.* **11**, 800.

Holden, P. (1955). Transmission of eastern equine encephalomyelitis in ring-necked pheasants. *Proc. Soc. exp. Biol. Med.* **88**, 607.

Hoogstraal, H. and Kaiser, M. N. (1961). Ticks from European birds migrating through Egypt into Africa. *Science, N.Y.* **133**, 277.

Hurlbut, H. S. (1950). The transmission of Japanese B encephalitis by mosquitoes after experimental hibernation. *Am. J. Hyg.* **51**, 265.

Johnson, K. M., Mackenzie, R. B., Webb, P. A. and Kuns, M. L. (1965). Chronic infection of rodents by Machupo virus. *Science, N.Y.* **150** (3703): 1618.

Karstad, Lars and Hanson, R. P. (1957). Vesicular stomatitis in deer. *Am. J. vet. Res.* **18**, 66.

Karstad, Lars (1961). Reptiles as possible reservoir hosts for Eastern encephalitis virus. *Trans. N. Am. Wildl. Conf.* **26**, 186.

Karstad, Lars, Spalatin, J. and Hanson, R. P. (1961). Natural and experimental infections with the virus of eastern encephalitis in wild rodents from Winconsin, Minnesota, Michigan and Georgia. *Zoonoses Res.* **1**, 87.

Kissling, R. E., Chamberlain, R. W., Sikes, R. K. and Eidson, M. E. (1954). Studies on the North American arthropod-borne encephalitides. III. Eastern equine encephalitis in wild birds. *Am. J. Hyg.* **60**, 251.

Kissling, R. E., Stamm, D. D., Chamberlain, R. W. and Sudia, W. D. (1960). Birds as winter hosts for Eastern and Western equine encephalomyelitis virus. *Am. J. Hyg.* **66**, 42.

Kozuch, O., Nosek, J., Ernek, E., Lichard, M. and Albrecht, P. (1963). Persistence of Tick-borne encephalitis virus in hibernating hedgehogs and door-mice. *Acta virol., Prague* **7**, 430.

La Motte, L. C. (1958). Japanese B encephalitis in bats during simulated hibernation. *Am. J. Hyg.* **67**, 101.

Lumsden, W. H. R., Williams, M. C. and Mason, P. J. (1961). A virus from insectivorous bats in Uganda. *Ann. trop. Med. Parasit.* **55**, 389.

McLean, D. M. and Larke, R. P. B. (1963). Powassan and Silverwater virus: ecology of two Ontario arboviruses. *Can. med. Ass. J.* **88**, 182.

Miles, J. A. R. (1960). Epidemiology of the arthropod-borne encephalitides. *Bull. Wld Hlth Org.* **22**, 339.

Nosek, J., Gresikova, M. and Rehacek, J. (1961). Persistence of tick-borne encephalitis virus in hibernating bats. *Acta virol., Prague* **5**, 112.

Reeves, W. C., Bellamy, R. E. and Scrivani, R. P. (1958). Relationships of mosquito vectors to winter survival of encephalitis viruses. I. Under natural conditions. *Am. J. Hyg.* **67**, 78.

Reeves, W. C. (1959). Problems of overwintering and natural maintenance of mosquito-borne viruses. *Int. Congr. trop. Med. Malar.* No. 6. **5**, 48.

Reeves, W. C. and Hammon, W. McD., in collaboration with Long-Shore, W. A., McClure, H. E. and Beib, A. F. (1962). Epidemiology of the arthropod-borne viral encephalitides in Kern County, California 1943–1952. *Univ. Calif. Publs publ. Hlth* **4**, 1–257.

Rudnick, A. (1966). Studies of the ecology of dengue in Malaysia. *Bull. Wld Hlth Org.* **35**, 78.

Shepherd, R. C. and Williams, M. C. (1964). Studies on viruses in East African bats (*Chiroptera*). I. Haemagglutination inhibition and circulation of arboviruses. *Zoonoses Res.* **3**, 125–139.

Scherer, W. F., in collaboration with Buescher, E. L., McClure, H. E., Moyer, J., Rosenberg, M. Z., Yoshii, M., Okada, Y., Izumi, T., Gresser, I., McCown, J., Okuno, J., Kitaoka, M., Ogata, T., Southam, C. M., Flemings, M. B., Noguchi, A., Hardy, J. L., and Bullock, H. R. (1959). Ecological studies of Japanese encephalitis virus in Japan, I to IX. *Am. J. trop. Med. Hyg.* **8**, 644–722.

Shope, R. E., Causey, C. E. and Causey, O. R. (1961). Itaqui virus, a new member of the arthropod-borne group C. *Am. J. trop. Med. Hyg.* **10**, 264.

Shope, R. E. and Whitman, L. (1966). Nepuyo virus, a new group C agent isolated in Trinidad and Brazil. II. Serological studies. *Am. J. trop. Med. Hyg.* **15**, 772.

Simkova, A. (1963). Quantitative study of experimental Tahyna virus infection in potential reservoir animals. *Acta virol., Prague* **7**, 414.

Simpson, D. I. H. (1966). The susceptibility of *Arvicanthis abyssinicus* (Ruppell) to infection with various arboviruses. *Trans. R. Soc. trop. Med. Hyg.* **60**, 248.

Smith, C. E. G. (1956). A virus resembling Russian spring-summer encephalitis virus from an *IXODID* tick in Malaya. *Nature, Lond.* **178**, 581.

Smith, C. E. G. (1958). The history of dengue in tropical Asia and its probable relationship to the mosquito *Aedes aegypti*. *J. trop. Med. Hyg.* **59**, 243.

Smith, C. E. G. (1962). Ticks and viruses. *Symp. zool. Soc. Lond.* No. 6, 199.

Smith, C. E. G., Varma, M. G. R. and McMahon, D. (1964). Isolation of Louping-ill virus from small mammals in Ayrshire, Scotland. *Nature, Lond.* **203**, 992.

Smith, C. E. G. (1968). Naming and grouping arboviruses: publication of a catalogue. *Ann. trop. Med. Parasit.* (in press).

Sulkin, S. E. (1961). Laboratory-acquired infections. *Bact. Rev.* **25**, 203.

Sulkin, S. E., Allen, R. and Sims, R. (1963). Studies of arthropod-borne virus infections in chiroptera. 1. Susceptibility of insectivorous species to experimental infection with Japanese B and St. Louis encephalitis viruses. *Am. J. trop. Med. Hyg.* **12**, 800.

Sulkin, S. E., Sims, R. and Allen, R. (1964). Studies of arthropod-borne virus infection in chiroptera. 2. Experiments with Japanese B and St. Louis encephalitis viruses in the gravid bat. Evidence of transplacental transmission. *Am. J. trop. Med. Hyg.* **13**, 475.

Svedmyr, A., von Zeipel, G., Borg, K. and Hansen, H. J. (1965). Infections with tick-borne encephalitis virus in the Swedish population of the Elk (*Alces a. alces*). *Acta path. microbiol. Scand.* **65**, 613.

Taylor, R. M., Hurlbut, H. S., Work, T. H., Kingston, J. R. and Frothingham, T. E. (1955). Sindbis virus: a newly recognised arthropod-transmitted virus. *Am. J. trop. Med. Hyg.* **4**, 844.

Taylor, R. M., Work, T. H., Hurlbut, H. S. and Rizk, F. (1956). A study of the ecology of West Nile virus in Egypt. *Am. J. trop. Med. Hyg.* **5**, 579.

Ten Broeck, C. (1938). Birds as possible carriers of the virus of equine encephalomyelitis. *Archs Path.* **25**, 759.

Thomas, L. A., Eklund, C. M. and Rush, W. A. (1958). Susceptibility of Garter snakes (*Thamnophis* spp.) to Western equine encephalomyelitis virus. *Proc. Soc. exp. Biol. Med.* **99**, 698.

Thomas, L. A. and Eklund, C. M. (1960). Overwintering of Western equine encephalomyelitis virus in experimentally infected garter snakes and transmission to mosquitoes. *Proc. Soc. exp. Biol. Med.* **105**, 52.

Van Tongeren, H. A. E. (1959). Central European encephalitis—epidemiology and vectors. *Int. Congr. trop. Med. Malar.* No. 6. **5**, 174.

Webb, H. E. (1965). Kyasanur Forest Disease Virus in three species of rodents. *Trans. R. Soc. trop. Med. Hyg.* **59**, 205.

Weinbren, M. P. and Mason, P. J. (1957). Rift Valley Fever in a wild field rat (*Arvicanthis abyssinicus*): a possible natural host. *S. Afr. med. J.* **31**, 427.

Williams, M. C., Simpson, D. I. H. and Shepherd, R. C. (1964). Studies on viruses in East African bats (*Chiroptera*). 2. Virus isolation. *Zoonoses Res.* **3**, 141.

W.H.O. (1967). Arboviruses and Human Disease. *Tech. Rep. Ser. Wld Hlth Org.* No. 369.

Work, T. H. (1958). Russian spring-summer virus in India. Kyasanur Forest disease. *Prog. med. Virol.* **1**, 248.

DISCUSSION

CAVILL: It is probable that birds are important maintenance hosts for the Group A arboviruses and probably also for some of the Group B arboviruses. To what extent are arboviruses capable of producing clinical disease in birds apart from Eastern encephalitis in pheasants?

How important is the red grouse as a maintenance host for louping-ill virus? Should not the names of "Eastern equine encephalitis" etc., be abandoned in favour of Eastern encephalitis, since the causal agents are not strictly equine viruses?

SIMPSON: I agree entirely with your last statement, I only included them because many still persist in calling them by these names. We know that Williams and others have shown that the red grouse is of some importance in the maintenance of louping-ill virus in this country but I think further work needs to be done on this. There is, as far as I know, no definite evidence of any infection with Group A viruses other than Eastern encephalitis in pheasants.

Symp. zool. Soc. Lond. (1968) No. 24, 29–38.

A RED SQUIRREL DISEASE

A. D. VIZOSO

Virus Research Unit, Medical Research Council Laboratories,
Woodmansterne Road, Carshalton, Surrey, England

SYNOPSIS

European red squirrels (*Sciurus vulgaris*) have a history of marked fluctuations in numbers and the British variety is no exception. During the early years of this century when the species became scarce or extinct in many districts, crude descriptions of some form of disease were a feature of many reports. Whatever the initiating factors, disease, not identified at that time, caused many deaths on a wide scale.

It was therefore interesting to learn of a further outbreak in this country (Edwards, 1962) in which symptoms, corresponding with earlier descriptions, were carefully observed but no common agent was identified from affected squirrels sent to various laboratories. However, the evidence suggested that a virus infection might well be responsible; this could easily have been overlooked in routine post-mortem examinations. Therefore an investigation was commenced to search for viral agents in wild populations of red squirrels.

As a preliminary step, an apparently healthy animal was examined and yielded viruses of two distinct characters—RS3 and RS3/2. Soon afterwards, the population from which this animal came suffered an outbreak of lethal disease with symptoms similar to those described by Edwards. From animals showing those symptoms, an agent of the type RS3/2 was isolated and designated RS6, the virus with which most of the subsequent studies were made. Whilst this work was going on, sick squirrels, found near the site of Edwards' initial outbreak, were sent for examination; these showed all the symptoms of the disease. Agents like RS6 were again isolated. Tests for other viral agents proved negative. Of the agents isolated from the red squirrels, one of them (RS3) was a virus of the encephalomyocarditis group differing from the main group in that it lacked the ability to agglutinate sheep red cells (Vizoso, Vizoso and Hay, 1964). The others, RS3/2 and RS6, have some of the characteristics of the group of myxoviruses and are capable of transforming baby hamster kidney cells (BHK) which, when inoculated again into suckling and adult hamsters, produce transplantable tumours (Vizoso, Hay and Battersby, 1966). During 21 passages in tissue culture the transformed BHK cells produced no detectable virus but inoculated into hamsters they produced tumours from which the virus could be recovered.

THE PROBLEM

European red squirrels (*Sciurus vulgaris*) have a history of marked fluctuations in numbers and the British variety, *S. vulgaris leucourus*, is no exception. The cycle has often been described as a seven- to eight-year period with a slow build-up leading to abundance, followed by a sharp decline. In Finland and in Britain this population decline has been associated with disease (Lampio, 1948; Middleton, 1930; Edwards, 1962 and Shorten, 1964).

During the late nineteenth and early twentieth century in Britain, red squirrels had become abundant in many areas and descriptions of disease and subsequent decline in numbers are well documented (Middleton, 1930; Shorten, 1954 and Vizoso, 1967.) Unfortunately, the descriptions are generally crude, relying more on comparison with other known diseases ("mange", "consumption", "coccidiosis", "diphtheria", "distemper") than upon any exact description of symptoms. Little if any effort was directed to pathological examination. Nevertheless, there is a similarity in descriptions during different years from widely separate districts which suggests the possibility of a common agent, endemic in the species.

There has been much speculation about the reasons for the decline of the native red squirrel of Britain since the beginning of this century. A popular theory is that the introduced American grey squirrel is mainly responsible: if not by direct slaughter, then by competition for food and nesting sites, by transmitting disease or by some less obvious stress factor. An alternative theory is that a reduction of the most suitable habitat for red squirrels has led to virtual replacement by the grey squirrel in many areas. Such a change in habitat need not be an obvious alteration in type or extent of woodland but could even be a change in degree of disturbance, the elimination of a seasonal food or some less evident factor.

A comparison of the known dates and places of introductions of grey squirrels with known dates and places of outbreaks of red squirrel disease neither proves nor disproves a connection (Shorten, 1964). Enzootics amongst red squirrels occurred where no grey squirrels are known to have existed before the outbreak but at this stage it would be impossible to confirm that no abortive introduction took place in the localities concerned. It could be argued that some of the outbreaks of disease show a connection with grey squirrel colonies spreading from introductions. Against this, there is no pattern of concentration of disease around such colonies but ordinary methods of collecting the evidence might have masked such a pattern. Perhaps the grey squirrel introduced a disease agent to the red squirrel to which it had itself become largely immune or perhaps the red squirrel population carried a widespread and regulated infection which maintained itself in apparently healthy animals between epidemics, becoming lethal in animals exposed to stress.

Whatever the initiating factors of population declines, unidentified epidemic disease caused many deaths amongst red squirrels in the first decades of this century. It was therefore interesting to learn of a further outbreak (Edwards, 1962) in 1960 in which carefully described symptoms tallied with those of earlier accounts. Again no common agent was identified although affected animals were sent to various laboratories. The evidence suggested a virus infection might be responsible. In a routine post-mortem examination an unfamiliar viral agent could easily be overlooked and it was decided to search for viral agents in wild populations of red squirrels.

There were two hypotheses: 1. that enzootics arose amongst red squirrels when stress or "disturbed habitat" activated a hitherto latent disease, carried in apparently healthy animals and common throughout populations, and 2. that an agent carried by immune grey squirrels caused disease when transmitted to red squirrels. As a preliminary step it seemed logical to investigate a population which had not yet been in contact with grey squirrels, following this, when possible, with similar work on red and grey squirrels living within one area. It is impossible to be sure that any population of grey squirrels in Britain has at all times avoided contact with red squirrels and a study of viral agents peculiar to incoming greys could perhaps best be conducted in their country of origin.

THE DISEASE

Disease affecting red squirrels in Finland during a sharp decline was said to have been coccidiosis (Lampio, 1948) but Middleton (1930) considered that the widespread epizootics which affected red squirrels between 1900–1925 in England, Wales, Scotland and Ireland produced symptoms in many cases unlike those of coccidiosis. Often a mange-like, scabby condition was described with discharge from the eyes and nose.

In 1960 an outbreak of disease in Shropshire virtually exterminated a community of red squirrels: only one animal escaped. Edwards (1962) gives this description of external symptoms:

"The course of the disease is normally about one week, and apparently the mortality rate approaches 100 per cent. The first symptom is a conjunctivitis starting as petechial haemorrhages on the rims of the eyelids, and later developing a tenacious whitish-grey discharge. . . . Shortly after this there is a nasal discharge, the lips become swollen, and the inside of the mouth becomes sore and ulcerated. The skin lesions are described as running sores of the lower jaw in the centre of the intermaxillary area. These were present in several cases. Three cases had ear lesions leading almost to decomposition of the concha, but no offensive odour was noticed. By the fifth day a throat swelling appears which causes difficulty in swallowing. The animal appears thirsty and attempts to drink without success. Finally, the squirrels become almost blind and very weak and the inability to eat makes death a certainty.''

The author makes the interesting observation that grey squirrels appeared in the locality for the first time about two months before the outbreak of disease and were in complete occupation two years later.

Although I did not examine any squirrels dying from this outbreak, two young red squirrels were sent from the same area later by Mrs. Beatrice Eatough and these showed similar symptoms. From them an agent has been isolated.

An outbreak of disease in 1963–66 amongst red squirrels at Thetford Chase in Norfolk was of particular interest since no grey squirrels had ever been observed in the locality. Some 30 dead animals were inspected and 3 which were sick. The symptoms tallied closely with those described by Edwards but throat swellings and ear lesions were not found. The animals appeared to be nearly blind and Forestry employees invariably described the diseased squirrels as having myxomatosis. Sick animals were easy to approach, finally running to the nearest tree and circling the trunk 5 or 6 feet above ground. Dead squirrels found during the daily search were usually on roads or rides and carried large numbers of ectoparasites.

Post-mortem examination showed red and grey hepatization of the lungs, occasional haemorrhagic areas in the kidneys, a possible hypertrophy of the suprarenal glands and very small spleens. Some squirrels also had haemorrhagic meninges. On histological examination necrotic foci were found in some livers but too few to be connected with certainty to the disease. No lesions were found in the pancreas, brain or spinal cord. Stomach and intestines were practically empty. The only extensive and invariable damage was to the lungs. From lung suspensions a viral agent was isolated.

THE AGENT

As a preliminary step, an apparently healthy red squirrel from Thetford was examined. From a small length of intestine and its contents a suspension was made and two viruses (RS3 and RS3/2) were isolated.

Virus RS3 was isolated in suckling mice and L-cell tissue culture. It proved to be ether-resistant and capable of producing plaques in mouse ascites cells when ribonucleic acid extracted from infected cells was inoculated. It was identified as encephalomyocarditis (EMC) but differed from the well known strains of this group in lacking the ability to agglutinate sheep red cells (Vizoso *et al.*, 1964). No detectable antibodies were present in the animal.

Virus RS3/2 was isolated after inoculation of the filtered gut suspension into ten 11-day-old chick embryos. Inoculations were made into the chorio-allantois or the amniotic cavity. Pocks of about 0·5 to 1·0 mm in diameter were found in the chorio-allantoic membrane of the infected eggs. The amniotic fluid contained small amounts of agglutinin for red cells of various mammalian species. It could be transferred serially in embrionated eggs by amniotic inoculation and killed about 50% of the embryos, many of which showed extensive haemorrhagic areas. RS3/2 proved to be ether sensitive. Whatever the route of inoculation, it produced no effects in suckling or adult mice, rabbits, hamsters or guinea-pigs. It produced cytopathic effects in tissue cultures of continuous lines, such as L-cell, mouse fibroblast, human HeLa cells and especially in baby hamster kidney (BHK) cells.

Shortly afterwards, a further outbreak of disease at Thetford yielded fresh material. From the lungs of squirrels showing symptoms like those described by Edwards (above) an agent of the RS3/2 type was isolated. This agent, called RS6, has been the subject of intensive study. Red squirrels from Shropshire, close to the district where Edwards observed the disease described in his paper, were also examined and an agent with all the characteristics of RS6 was again isolated from lung suspension here (Vizoso *et al.*, 1966).

Samples of lung suspension purified by calcium phosphate chromatography and by sucrose gradients were examined under a low resolution electron microscope. Negative stained preparations showed some pleomorphic forms similar to those described by Almeida and Tyrrell (1967) for some uncharacterized human respiratory viruses grown in human organ culture. Some of the particles resembled those of Newcastle virus disease (NVD). Some electron negative particles could also be found, one of which showed an outer membrane with an internal helical component wound around an electron dense nucleoid. Whether one specific agent is represented by such a form is not known. The sizes vary from 200 to 500 mμ. Some electron dense nucleoids surrounded by a thin membrane were also observed in virus which had been obtained from tumours induced in hamsters. The bacteriologically sterile lung suspension produced very marked cytopathic effects in the cell lines listed above. Diluted suspensions produced destruction of the cells and further tubes could be inoculated with the supernatant of the destroyed cultures. After 4 to 5 passages the cytopathic effect disappeared.

At greater dilutions both RS3/2 and RS6 produced on inoculation

D

an effect reminiscent of that caused by Rous sarcoma virus or rabbit fibroma in tissue culture. Foci, or aggregates of the small cells, could be seen in the cultures, the tubes being specked with mounds of cells with blank areas in between. Haemadsorption could be demonstrated easily in tissue culture with lung suspension. As yet we have failed to produce plaques in agar-covered cell cultures and, lacking a quantitative assay method, have to rely mainly upon the haemagglutinating activity of the agent.

The agent can be purified by treatment with Arcton and can be absorbed and eluted from calcium phosphate columns (Taverne, Marshall and Fulton, 1958). It elutes at a phosphate molarity of between 0·4–0·5M and can be further purified in sucrose gradients, sedimenting to about 30–40% sucrose in a broad band identified by haemagglutinin activity and by infectivity. When the band containing the agent was again subjected to a density gradient, in caesium chloride, it formed another broad band with an average buoyant density of 1·21. Haemagglutinin could still be detected but infectivity was lost.

At 37°C the agent RS6 showed some haemolytic activity which was lost after a single passage in embrionated egg. Quite the reverse was found by Matsumoto and Maemo (1962) for Sendai virus. It is probable that in both cases the haemolytic activity of the virus is also responsible for cell fusion.

One of the characteristics of this agent is the ability to transform baby hamster kidney cells (Vizoso et al., 1966). The observed criteria of cell transformation may be summarized as:

a) the change in cell morphology,
b) the loss of contact inhibition,
c) the ability of cells to grow into colonies in soft agar as described by Macpherson and Montagnier (1964),
d) the possible ability of cells to form intercellular bridges across which deoxyribonucleic acid may be transferred (Bendich, Vizoso and Harris, 1967),
e) the ability of transformed cells to grow into tumours when inoculated into suckling or adult hamsters.

Many transformed cells exhibit eosinophilic and basophilic inclusions in the cytoplasm after several passages. Giant cells were also produced with large nuclei which appear to break into micronuclei often surrounding the eosinophilic area. These giant cells have large vacuoles in the cytoplasm and appear to be viable as far as can be judged by staining cultures with haematoxylin and eosin Fig. (1). The first passage of transformed cells showed chromosome changes including deranged

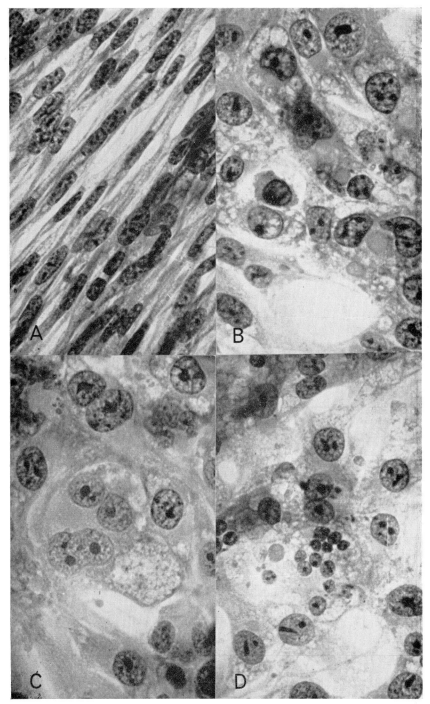

FIG. 1. Baby hamster kidney cells in cover slip cultures, stained with haematoxylin and eosin. A, normal cells. B, C and D, transformed cells. B, transformed cells showing enlarged and deformed nuclei and eosinophilic inclusion bodies in the cytoplasm. C, multinucleated cells containing an eosinophilic inclusion and cytoplasmic vacuoles. D, giant cells with micronuclei.

mitosis, trailing and poliploidy. Especially on isolation from lung material, cell fusion occurred.

The characteristics of the agent suggest a virus of the myxovirus group.

It is interesting that the transformed cells produced by this agent do not yield any detectable virus into the supernatant, nor can any be recovered from the cells after disruption by sonic treatment or freezing and thawing. However, after some 20 passages in tissue culture, they are capable of inducing transplantable tumours in hamsters and from some of these tumours the agent can be recovered.

It could be argued that the cell lines used in these studies have become carriers. This seems especially true of the baby hamster kidney cells. The majority, after a number of passages, still carry the eosinophilic inclusion bodies, the basophilic inclusions having become very rare. We could not test the possibility that such eosinophilic inclusions in the cytoplasm correspond to viral antigen, since we failed to prepare anti-viral sera; however, Miyamoto, Homma and Ishida (1967) describe what appears to be a similar situation for haemadsorption type 2 virus. Here the cytoplasmic inclusion bodies of each carrier cell contained nucleocapsid material. The virus-cell relationship described by these authors was "a perfect relationship between viral and host genomes, so that while each cell is infected, yet every one is capable of transmitting the infection and of forming new clones of cells. It remains to be determined whether there is a direct genetic interaction between host and viral genomes."

In the case of RS6 and baby hamster kidney cells the changes in morphology of the cells, the cloning efficiency in soft agar and the whole process of cell transformation suggests a genetic interaction between host and virus.

DISCUSSION

The fact that this virus, like some of the myxovirus group (Miyamoto et al., 1967 and Walker and Hinze, 1962), can exist in the carrier state in some susceptible cells or survive without destroying the cells while continuing to produce small amounts of virus (Choppin, 1964) and under certain conditions may then produce overt disease, arouses speculation as to its rôle in the host population.

A community of squirrels, dependent on a staple diet of tree seed which fluctuates in abundance, can in turn be expected to undergo fluctuations in numbers. If food abundance follows a rough cycle, this may be converted through breeding response into a more or less regular cycle in squirrel abundance.

Such a cycle occurs and is typified by a slow rise followed by a sharp fall in numbers. At the peak, overcrowding is likely to cause stress, stress may provoke the appearance of overt disease and mortality from disease may then remove overcrowding and stress. The virus could then be said to act as a population controller.

The more or less regular pattern of the cycle could be broken by changes in staple food, addition of new sources of stress or a change in the nature of the disease. In the case of the British red squirrel, the upswing of the cycle may, in many districts, have been inhibited by destruction or disturbance of woodland during the twentieth century. New stress factors, including the presence of the American grey squirrel, may have led to the expression of overt disease even in sparse populations—and, eventually, to local extinction.

ACKNOWLEDGEMENTS

The author wishes to thank all those who helped to provide information and material, notably Mrs. Beatrice Eatough and Mr. Desmond Green. Mr. T. Battersby gave valuable help in tissue culture preparations.

REFERENCES

Almeida, J. D. and Tyrrell, D. A. J. (1967). The morphology of three previously uncharacterized human respiratory viruses that grow in organ cultures. *J. gen. Virol.* **1**, 175–178.

Bendich, A., Vizoso, A. D. and Harris, R. G. (1967). Intracellular bridges between mammalian cells in culture. *Proc. natn Acad. Sci. U.S.A.* **57**, 1029–1035.

Choppin, P. W. (1964). Multiplication of a myxovirus (SV5) with minimal cytopathic effects and without interference. *Virology* **23**, 224–233.

Edwards, F. B. (1962). Red squirrel disease. *Vet. Rec.* **74**, 739–41.

Lampio, T. (1948). Squirrel economy in Finland based on natural prerequisites. *Suom. Riista* **2**, 9–147.

Macpherson, I. and Montagnier, L. (1964). Agar suspension culture for the selective assay of cells transformed by polyoma virus. *Virology* **23**, 291–294.

Matsumoto, T. and Maemo, K. (1962). A host induced modification of hemagglutinating virus of Japan (HVJ, Sendai virus) in its haemolytic and cytopathic activity. *Virology* **17**, 563–570.

Middleton, A. D. (1930). The ecology of the American grey squirrel in the British Isles. *Proc. zool. Soc. Lond.* 809–843.

Miyamoto, T., Homma, M. and Ishida, N. (1967). Infectivity expression of a carrier cell line HeLa/HA2. *Virology* **33**, 591–597.

Shorten, M. R. (1954). Squirrels. Collins New Naturalist, London.

Shorten, M. R. (1964). Introduced menace. *The Journal of the American Museum of Natural History* **73**, 42–49.

Taverne, J., Marshall, J. H. and Fulton, F. (1958). The purification and concentration of viruses and virus soluble antigens on calcium phosphate. *J. gen. Microbiol.* **19**, 451–461.

Vizoso, A. D., Hay, R. and Battersby, T. (1966). Isolation of unidentified agents capable of morphologically transforming hamster cells in vitro. *Nature, Lond.* **209**, 1263–1264.

Vizoso, A. D., Vizoso, M. R. and Hay, R. (1964). Isolation of a virus resembling encephalomyocarditis from a red squirrel. *Nature, Lond.* **201**, 849–850.

Vizoso, M. R. (1967). Squirrel populations and their control. Forestry Supplement, 14–20.

Walker, D. L. and Hinze, H. C. (1962). A carrier state of mumps virus in human conjunctiva cells. *J. exp. Med.* **116**, 739–750.

DISCUSSION

GREEN: Were either of the red squirrel agents detected in other rodents which shared the same environment during outbreaks of the disease?

VIZOSO: We did not examine other species in the locality during a disease outbreak amongst red squirrels; we have found a similar agent, as yet unidentified, in grey squirrels, but in different localities.

CHU: Dr. Vizoso mentioned that one of the viruses he isolated from the lungs of red squirrels looked similar to the human respiratory virus recently isolated by Dr. Tyrell using organ cultures. Does he refer to the virus which was shown by Dr. Tyrell to be morphologically similar to the avian bronchitis virus?

VIZOSO: Yes: the morphology of some of the particles found in lung suspensions from diseased red squirrels is similar to the human respiratory viruses 229E and B814 grown in organ cultures (which Almeida and Tyrell described as resembling those of infectious bronchitis of chickens).

Symp. zool. Soc. Lond. (1968) No. 24, 39–49

ULCERATIVE DERMAL NECROSIS OF SALMONIDS IN IRELAND

J. T. CARBERY

Veterinary Research Laboratory,
Abbotstown, Castleknock, Co. Dublin, Eire

SYNOPSIS

Ulcerative dermal necrosis, an infectious disease of some salmonids, is described. The primary lesion is a necrosis of the skin proceeding to ulceration of the dermis. Various secondary invaders may complicate the clinical and post-mortem findings. Some of the epizootiological aspects are dealt with. The disease in Ireland is confined to certain species of the genus *Salmo*—mature *Salmo salar* and *S. trutta* only are affected. The incidence of infection is greatest in the colder seasons of the year, falling very markedly in summer. The differential diagnosis is concerned with fungus infection and furunculosis. The aetiological agent has not been isolated although K. L. Strickland and J. T. Carbery (1968, unpublished) showed the disease was infectious and that it could be transmitted through filters of 200 mμ porosity.

INTRODUCTION

Ulcerative dermal necrosis (UDN) is an infectious disease of Atlantic salmon, sea-trout and brown trout. The present epizootic was first reported from the south west of Ireland in late 1964. A clinically similar disease was seen in Britain from 1877–1913 (Buckland and Walpole, 1880; Salmon Fisheries Report, 1882; Hume Patterson, 1903; Malloch, 1910; Grimble, 1913).

DESCRIPTION

Pathogenesis and external lesions

In describing the pathogenesis of UDN, knowledge obtained experimentally (K. L. Strickland and J. T. Carbery, unpublished) is being combined with that obtained from the study of the naturally occurring disease (Carbery and Strickland, 1968a). The initial lesions develop on the unscaled areas of the skin as one or more small necrotic— hence white—spots. The lesion extends and, in some cases, at this stage has the feeling of a shallow blister. The necrosis proceeds to deep ulceration, involving the dermis. Fijian (1966) describes a somewhat similar process in the experimental reproduction of haemorrhagic septicaemia (infectious dropsy) of carp. Sooner or later the necrotic area or the ulcer may become infected with species of fungus, chiefly *Saprolegnia*. In quite a number of natural outbreaks fungal invasion either does not take place, or occurs only late in the disease; in others, e.g. in many of the rivers of Cork, Kerry and Waterford, fungal invasion takes place at a very early stage and its development parallels that of the primary lesion. *Saprolegnia* in many cases plays a very important part in the termination of the disease. A number of other secondary invaders may contribute to the death of a fish sick with UDN. *Aeromonas liquefaciens* and various species of *Pseudomonas*—mainly *Ps. fluorescens* —alone or in combination with one another or with *Saprolegnia* usually are involved in the final stages of the disease. A terminal generalized infection of *A. liquefaciens* or *Ps. fluorescens* will cause the superimposition of septicaemic symptoms on those of UDN. *Saprolegnia* is the commonest and most important of the secondary invaders. In many instances it has been possible to keep UDN affected fish alive, and to spawn them successfully, by treating them with malachite green thus high-lighting the important part that the fungus plays in the death of the fish.

Internal lesions

Fish suffering from UDN appear normal internally on gross post-mortem examination except in cases where bacterial invasion has occurred. A generalized infection with *A. liquefaciens* or *Ps. fluorescens* will result in lesions of septicaemia being found.

EPIZOOTIOLOGICAL ASPECTS

Species affected

In our experience only members of the genus *Salmo* are affected clinically with UDN. The disease has been reported from a wide variety of freshwater fish. It is likely that confusion of a fungus infection with UDN has led to these reports. We have been able to infect *S. salar* and *S. trutta* (K. L. Strickland and J. T. Carbery, unpublished) but not *S. gairdnerii* (Carbery and Strickland, 1968b) with UDN. Furthermore, rainbow trout exposed to infection did not show any evidence of being carriers of UDN (Carbery and Strickland, 1968b).

Age group affected

In Ireland the only salmon which develop clinical symptoms of UDN are those which have been to sea and have returned to fresh water. Fry, parr, smolts and salmon in the sea show no signs of the disease. Age specificity would appear to apply to the disease in trout also. When trout are affected they are almost invariably fairly large specimens. An outbreak of the disease in brown trout in a river which contained no anadromous salmonids—due to an series of impassible weirs—affected the larger fish (over 1½ lbs) almost exclusively.

Seasonal variation

Morbidity rates are highest in late autumn, winter and spring when water temperatures are lowest. In summer the incidence falls so that the disease disappears almost entirely from most affected rivers. It would thus appear that the condition is temperature dependent, a not uncommon feature of many fish diseases reported in the literature (Ghittino, 1965; Pacheco and Guimaraes, 1933; Ross, Pelnar and Rucker, 1960; Rasmussen, 1965; Jensen, 1965a).

DIFFERENTIAL DIAGNOSIS

A diagnosis of UDN is at present based on clinical findings. Two conditions which may be confused with it are fungus infection and furunculosis.

Fungus infection

Various species of the fungus *Saprolegnia* may act as primary and lethal parasites of salmon whose resistance has been lowered. Kelts, due to their debilitated condition, are common victims of this organism. Fresh-run salmon caught and put into holding ponds often develop fungus infection due to the stress and superficial abrasion consequent on handling. Fungus infection is therefore likely to cause confusion in three types of fish—kelts, salmon held for stripping and experimental fish.

In fish held in captivity the pattern of development of the infection can be observed. In primary fungus infection, in contrast to the secondary infection found in UDN, growth commences at various sites all over the body. Rings or portions of rings of fungal growth develop which gradually fill in to give solid patches which may coalesce to form larger areas. In our experience ulcers involving the dermis such as are found in UDN are never seen in simple fungus infection. Failure to differentiate between UDN and primary fungus infection could lead to confusion in experiments designed to elucidate the aetiology of UDN.

Furunculosis

Furunculosis in some of its forms may be confused with UDN. Where the so-called furuncles develop on the head and rupture, the resultant lesions may resemble the later stages of ulceration found in UDN. The growth of fungus may further confuse the issue.

The isolation of *Aeromonas salmonicida* from the lesions and/or internal organs allows a diagnosis of furunculosis to be made. The fact that furunculosis is mainly a disease of summer months is also of help.

AETIOLOGY

Hume Patterson (1903) studied the bacteriology of the earlier similar disease. He conducted infectivity trials as a result of which he stated that the cause of the disease was not a *Saprolegnia*, as Huxley (1882) had thought, but a bacterium which he called *Bacillus salmonis pestis*.

Bissett (1946) identified a culture of *B. salmonis pestis* as a mixture of a *Proteus* and *Pseudomonas fluorescens*. *Aeromonas liquefaciens* has been classified in the genus *Proteus* (Bullock, 1964; Snieszko and Bullock, 1962) and *Proteus hydrophilus* is sometimes used as a synonym. Consequently, the flora that Hume Patterson isolated was probably a mixture of *A. liquefaciens* (syn. *A. hydrophila*, *A. punctata*) and *Ps. fluorescens*.

Jensen (1965b) tentatively diagnosed the present condition as a "coldwater form of columnaris" but emphasized that his diagnosis would have to be confirmed by experimental work. Brown (1966) and Brown and Collins (1966) later formed the opinion that the condition was columnaris disease. White (1965, personal communication) of this laboratory attempted to confirm Jensen's diagnosis but could not do so.

Early involvement of *Saprolegnia* in the disease has led some workers (Stuart and Fuller, 1968) to state that the fungus may be the primary cause.

We began work on UDN in 1966 and in that year examined more than 100 salmon affected with the disease using freshly-killed salmon only. We found (Carbery and Strickland, 1968a) no species of bacterium to be in constant association with the disease—confirming a previous finding of D. S. White, personal communication—and in fact found the internal organs of most freshly-killed cases of UDN to be sterile on the media used. Where bacteria were involved, the most common organisms isolated from internal organs were *A. liquefaciens* and *Ps. fluorescens*. *Myxobacteria* were isolated from the external surfaces of 30% of diseased fish. They were isolated also from 36·5% of healthy fish.

In September, 1966 we transmitted the disease to a healthy salmon from an affected one using a simple syphoning technique (K. L. Strickland and J. T. Carbery, unpublished). This experiment was successfully repeated seven times, showing that the disease was infectious and could be transmitted through the water, direct contact or contagion not being necessary. The disease was reproduced in healthy salmon by intraperitoneal injection of homogenates of tissues from diseased salmon which had been treated in various ways (K. L. Strickland and J. T. Carbery, unpublished). Inoculation of antibiotic treated homogenates and of filtered homogenates gave an infectivity rate of about 30%. The results of the experiments indicated that the size of the aetiological agent is less than 200 mμ.

SUMMARY

Ulcerative dermal necrosis, an infectious disease of some salmonids is described. Some of the epizootiological aspects are dealt with and the differential diagnosis is briefly treated. Studies into the aetiology are reviewed and the experimental work is outlined.

REFERENCES

Bisset, K. A. (1946). The Effect of Temperature on Non-Specific Infections of Fish. *J. Path. Bact.* **58**, 251–158.

Brown, Margaret E. (1966). Irish Salmon Disease. *Atlant. Salmon J.* **Summer 1966**.

Brown, Margaret E. and Collins, Vera G. (1966). Irish Salmon Disease—An Interim Report. *Salm. Trout Mag.* **178**, 180–188.

Buckland, F. and Walpole, S. (1880). Report on the Disease which has recently prevailed among the Salmon on the Tweed, Eden and other rivers in England and Scotland. H.M.S.O. London. **1880**.

Bullock, G. L. (1964). Pseudomonales as Fish Pathogens. *Devs. ind. Microbiol.* **5**, 101–108.

Carbery, J. T. and Strickland, K. L. (1968a). Ulcerative Dermal Necrosis. *Irish vet. J.* **22**, 171–175.

Carbery, J. T. and Strickland, K. L. (1968b). Resistance of Rainbow Trout to Ulcerative Dermal Necrosis. *Nature, Lond.* **217**, 1158.

Fijian, Nikola N. (1966). Experimental Transmission of Infectious Dropsy of Carp, *Bull. Off int. Épizoot.* **65**, 731–738.

Ghittino, P. (1965). Viral Hemorrhagic Septicaemia (VHS) in Rainbow Trout in Italy. *Ann. N.Y. Acad. Sci.* **126**, 468–477.

Grimble, A. (1913). *The Salmon Rivers of England and Wales (2nd Edition).* Kegan Paul, French, Trubner & Co. Ltd., London.

Hume Patterson, J. (1903). *On the Course of the Salmon Disease—a Bacteriological Investigation.* H.M.S.O. Glasgow.

Huxley, T. H. (1882). Q. Jl microsc. Sci. **22**, 311.

Jensen, M. H. (1965a). Research on the Virus of Egtved Disease. Ann. N.Y. Acad. Sci. **126**, 422–426.

Jensen, M. H. (1965b). Disease among Salmon in Irish Rivers 1964–1965. Report to the Minister for Agriculture & Fisheries. Dublin, **1965**.

Malloch, P. D. (1910). Life History and Habits of the Salmon, Sea Trout and other freshwater fish. A. & C. Black, London, **1910**.

Pacheco, G. and Guimaraes, V.R. (1933). Ichtyozooties dans les eaux bluxiales de l'Etat de sao Paulo. C.r. Séanc Soc. Biol. **114**, 1401–1404.

Rasmussen, C. J. (1965). A Biological Study of the Egtved Disease (INUL). Ann. N.Y. Acad. Sc. **126**, 422–426.

Ross, A. J., Pelnar, J. and Rucker, R. R. (1960). A virus-like disease of chinook salmon. Trans. Am. Fish. Soc. **89**, 160–163.

Salmon Fisheries (England and Wales), 21st Annual Report of the Inspectors of Fisheries (for the year 1881), H.M.S.O. London, **1882**.

Snieszko, S. F. and Bullock, G. L. (1962). Freshwater Fish Diseases caused by Bacteria belonging to the Genera Aeromonas and Pseudomonas. Fishery Leafl. Fish Wildl. Serv. U.S. No. 459.

Stuart, Mary R. and Fuller, Hubert T. (1968). Mycological Aspects of Diseased Atlantic Salmon. Nature, Lond. **217**, 90–92.

DISCUSSION

AUSTWICK: Do the fungal lesions ever heal naturally or do all the fish affected by fungi die?

CARBERY: We have evidence that if some fish go back down stream to the sea the action of the salt water will get rid of the fungus and the fish will start to come back up the river again. I do think that normally once a heavy fungus infection is established the fish will die from this. The fungus is responsible for the death of the fish perhaps in association with the other secondary invaders I have mentioned.

ROBERTS: In Ireland it was demonstrated that there was a very marked depreciation in the total amount of serum protein in these diseased salmon compared with the normal; we have also demonstrated this in Glasgow— it is down to about half the normal. We have taken this work on a little further and it does seem that the salmon is producing more globulin than the normal when it is diseased. This occurs whether the salmon is badly infected with fungus or not—it occurs in the early stages as well as in the late stages. Presumably, also, there will be an antibody reaction to the fungus. One thing I would like to ask Mr. Carbery: Our histological examinations agree completely with his—there is no obvious lesion whatsoever in the internal organs of the fish; the skin and subcutaneous tissue just down into the upper layers of the muscle is all that we have been able to find with evidence of lesions but, on a few of our fish (which it is believed had just come into the river), we found very small white spots with no evidence of ulceration. It has been suggested by our histologists that these spots might indicate a healing lesion. What do you think about the possibility that this infection might be obtained at sea and the ones that have not yet healed by the time they get to the river are, in fact, the ones that go down with U.D.N.? This is of course just a theory but the histology did, in fact, bear a resemblance to healing rather than degeneration.

CARBERY: It is not beyond the realms of possibility that the fish are actually infected in the sea and that the fish coming back to the rivers have partially recovered; they are, however, carrying the virus or, let us say, the agent and when they are under stress, as indeed they are in the change from salt to fresh water, this latent infection flares up and the clinical signs of U.D.N. appear. Our experience is slightly different from yours in that we have had no infected fish from the sea. One interesting point is that we had an outbreak in brown trout in a river in which there were no migratory fish, i.e. salmon or sea trout. The interesting question is, how did it get in? It might have been eels because they can cross wiers but we have never seen it in eels.

AVERY: Could this agent be carried by predators of the salmon because predators, such as otters, might tend to eat diseased fish.
   I wonder if they could carry it through to the brown trout in a situation where there was no direct contact with the sea?

CARBERY: I think the possibilities of how it gets into the river are limitless and until we know what the agent is we are doing a lot of speculation. We have not so far done any work on otters.

MUNRO: At the Marine Laboratories in Aberdeen, we have received a small number of fish from Sea Fisheries which are possibly 10 to 20 miles away from the nearest river; these fish have the symptoms Mr. Carbery showed on his slides. Where these fish came from we did not know but there are many possibilities as to where the fish could have picked up the disease.

HODKINSON: Have you examined any fish from rivers other than Irish rivers, because on the Lune in Lancashire they have had a tremendous outbreak of disease, loosing many fish prior to spawning. They don't get the ulcerated lesions you showed on the slides, in fact, it seems to me to be mainly fungal development; a haemorrhage is present but there is no ulceration associated with the skin.

CARBERY: I have not seen any fish apart from those from Irish rivers. In fish, fungus infection will become established on any debilitated tissues. The fungus will grow on any small necrotic spot and once it gains a foothold, it spreads rapidly. Sometimes the fungus moves in when there is extensive ulceration, in other rivers the infection with fungus appears at a very early stage in the disease and it may be that the time of invasion of the lesion depends upon the actual strains of fungus in the river, if indeed there are different strains. In quite a lot of rivers, for example, the Blackwater, the fungus moves in very early before ulceration has commenced.

HODKINSON: So how would you go about diagnosing this? We have not noticed the white blistery patches, we just seem to get masses of fungus on the dorsal surface.

CARBERY: If there are a lot of fish affected, it is probably U.D.N. Maybe the fish in some rivers are simply debilitated by the initial infection and at this stage the fungus moves in. I am surprised that no fish have been found with ulceration.

HODKINSON: I am not saying none—the majority have not. When ulceration occurs you seem to get this on the ventral surface and the fungal colonization mainly on the dorsal; there seems to be a possibility of a distinction between the two. There have been very few with these ventral ulcerations. The death of the fish seems to be mainly associated with the fungal development on the dorsal surface.

CARBERY: As regards ulceration, we feel this is typically on unscaled areas rather than on scaled areas.

HODKINSON: It does definitely seem to start on the head, on the base of the dorsal fin and spreads out. We don't get the ringworm development you showed but it definitely spreads out from the central area.

CARBERY: I would imagine that, if you could get these fish very early on before the fungus develops, you would probably find necrotic areas beneath where the fungus subsequently develops.

HODKINSON: Histological investigations are now proceeding.

SPILLING: How important is this disease from an economic viewpoint?

CARBERY: You can answer that in very many ways; as regards the commercial fisheries, it may be a long time before the effects are felt—in fact, there might be no effect at all as we reckon that about 10% of the fish are affected and I do not feel that it is really a serious threat to the salmon population. A lot of the commercial value of the salmon, certainly in Ireland and elsewhere, lies in its attraction for anglers. In this way the disease is very important economically and has already done a lot of damage to the hotel trade, the letting of fishings and the employment of ghillies and so on.

SPILLING: Why is the causative agent of this disease not known and what techniques are being used to try and identify it?

CARBERY: I cannot explain why it is not known; we have been looking for it but at the moment we are really concentrating on tissue culture. Indications are that it is probably a virus. We have been using primary cell lines from salmon smolts but so far we haven't isolated anything definite in them.

JONES: On the economic side salmon in Scotland are said to be worth £1,000,000 a year. I don't know how this figure is derived, whether it is based on the number of fish sold or if it is concerned with the ancillary industries. Obviously your investigation is well worthwhile.

CHAIRMAN'S SUMMING-UP

Considerable interest was shown in Dr. Brooksby's paper about the rôle of wild game with regard to foot-and-mouth disease particularly in this country. I think the best example one can cite with regard to this situation is what happened in Mexico from 1946 until 1950. A very severe epidemic occurred which was finally eradicated from domesticated animals by a combination of vaccination and slaughter. In the infected areas there was certainly a considerable population of wild animals, presumed susceptible, mainly deer and wild pig. No action was taken with regard to their possible rôle in the dissemination and persistence of infection. In spite of this the disease was eradicated. One feels, therefore, that where there is not an abundant wild game population, the importance of wildlife may not be so very great. It is, of course, a very different story in Africa, as Dr. Brooksby explained, with very large populations of wild game.

One of the things I liked about Dr. Simpson's paper on arboviruses, having had much personal interest in vesicular stomatitis over many years, the epidemiology of which is still very much in doubt, is that Dr. Gordon Smith's recommendation for the definition of an arbovirus would allow vesicular stomatitis to fall very happily into this group.

I take it from Dr. Vizoso's paper that, although he has not extended his investigations to any great extent, his reference to "similar agents" in the grey squirrel, suggests an inter-play between disease or inapparent infection in the grey squirrel which may yet have some bearing on the red squirrel population.

To sum up the last paper of this section I would like to echo the feeling that all of us must have of the great encouragement it is to listen to a paper which shows very clearly the excellence of the work that is going on in the investigation of a disease of fish such as U.D.N. The discussion it stimulated from people from various parts of Britain showed the attention, I would say deserved attention, which is now being given to an area of disease investigation which has so lacked it in the past.

BACTERIA

CHAIRMAN: C. L. OAKLEY

CHAIRMAN'S INTRODUCTION

It seems to me that most of these problems of infection in wild animals and their association with man involve very considerable behaviouristic studies; for instance, we might consider the degree to which animals are close to or separated from one another at different times in their life cycles and, of course, we might particularly consider the astonishing effect man exerts on the animals about him. Man is a remarkable and very curious animal; I am quite certain that if any of the animals we are considering in this symposium behaved as man does, we should visit upon them our continuous disapprobation Man has a great habit of doing large-scale ecological experiments and claiming to be immune from their natural results. He tends to associate himself in large masses in crowded, often grossly overcrowded, associations. He sets up enormous quantities of food stores and then complains that rats and people eat them, he has a continuous demand for justice and when he doesn't get it he supposes as a rule that someone is taking it away from him. All these things taken together make a fascinating study and perhaps persons who are concerned with the effects on man of disease-spreading animals pay far too little attention to them.

Beside all this, man, like other animals, keeps pets but he doesn't keep them for the same reasons most animals do—to get practical results from them—he does it to demonstrate his natural virtues; in consequence he is liable to complain when his treatment of these pets gives rise to disease in his own community. Anybody who is familiar with the way in which carrier pigeons, for instance, are transported across considerable areas of Europe in slatted boxes from which the agent of ornithosis is rapidly deposited all over the region associated with the railway ride will hardly be surprised to find that ornithosis is not particularly uncommon in human populations.

Symp. zool. Soc. Lond. (1968) No. 24, 51–73

SALMONELLA IN WILD ANIMALS

JOAN TAYLOR

Salmonella Reference Laboratory, Central Public Health Laboratory, Colindale, London, England

SYNOPSIS

A survey of the literature suggests that salmonellae have been isolated from practically all living animals; where such isolations have not been recorded it is probable that the animal has not been investigated. Such isolations have been made from animals in all parts of the world. The main animal groups will be discussed, insects, crustaceans, molluscs, fish, reptiles, amphibians, birds and mammals, and figures are given for the salmonella types from wild animals identified in the Salmonella Reference Library, 1951–1967. Comparisons will be made between the salmonella types found in various parts of the world.

The effect of the environment, feeding habits and natural behaviour in relation to infection will be discussed.

INTRODUCTION

The title of this paper suggests a vast subject and even a casual perusal of the literature shows this to be true. It would seem that salmonella have been isolated from many living creatures, in fact so many have been recorded that it seems reasonable to suppose that the absence of a record means that the particular creature may not have been studied. On the other hand it is possible to over-emphasize the importance of the isolation of salmonella from a species and to neglect the lesson to be learned from a related species from which salmonella have been rarely, if ever, isolated. For example, salmonella have not been isolated from grasshoppers (Lee and Mackerras, 1955) as far as I know, yet a comparison between those living on vegetation and carnivorous types might show very different results. When considering this subject, I realized that I had not come across any salmonella isolations from worms but, knowing that they were present in infected soil, I suspect that worms might be infected had they been tested. The title of this communication limits this subject to wild animals.

DEFINITION OF "WILD"

I have accepted as "wild" those animals living in game reserves but I have tried to omit animals in zoos. This means that cultures which I have received from exotic animals in this country have been omitted from my records but from foreign countries they have been included in spite of the fact that I may have had insufficient data to be dogmatic on this point. I found myself in a certain amount of difficulty when trying to assess the position of such birds as pheasants, as these may be reared domestically in early life, then be freed and come into the category of wild birds. Nevertheless, I have included pheasants and similarly reared animals as "wild" for the purpose of this paper.

BACTERIOLOGY

An organism is included in the salmonella group provided that it gives the essential biochemical reactions accepted by the Enterobacteriaceae Subcommittee of the International Association of Microbiological Societies (Report, 1958). Further biochemical reactions are then carried out which subdivide the group into subgenera I, II, III (Arizona) and IV (Kauffmann, 1966). The salmonella group, which includes all the subgenera, is divided into about 1800 members or, in scientific language, 1800 serotypes, by antigenic analysis. This is a technical procedure by which the body and the flagella of the organism are identified; when these are known the organism is named. Originally all members of subgenera I, II and IV were assigned names but since 1967 subgenus II new types have been un-named. Subgenus III, originally known as the Arizona group and still usually referred to by this name, was known only by the antigenic structure. For example, an organism giving the biochemical reactions of the salmonella group is found on antigenic analysis to have the somatic (body) structure 6, 8, the flagella are found to be of two varieties indicated by the symbols e,h and 1,2, then the full structure is written as 6,8 : e,h : 1,2; this organism is known as *Salmonella newport*, the further biochemical reactions are subgenus I. A member of subgenus III, the Arizona group, would give the necessary biochemical reactions and would be written Ar. 1,4 : 24–38, indicating that the body is the Arizona type 1,4 and that the two flagella types are Arizona 24 and 38. The salmonella subgenera I, II and IV were studied separately from the Arizona and only recently has it been agreed that all belong to the same group— the salmonella. The differentiation into subgenera is of importance in epidemiology, as the vast majority of salmonella isolated from sick animals are subgenus I. As will be shown later, salmonella subgenus II are most commonly found in reptiles; this is also true of subgenus III (Arizona) which have not been recorded in this paper. Arizona have been isolated from sick birds, although mainly domestic birds, in the

U.S.A. Subgenus IV is extremely rare so that it is not possible to relate it to disease or to any particular animal species. My own records for 1967 show that reptiles provided all but five strains of subgenus III and IV isolated from animals.

During the years 1951–1967 I examined strains from the following countries, Australia, Canada, Egypt, Germany, Ghana, Greece, Guyana, Hong Kong, India, Iran, Israel, Jamaica, Kenya, Malaya, Mauritius, New Zealand, Nigeria, Norway, Seychelles, Sierra Leone, South Africa, Tanzania, Turkey, Uganda, U.S.A., Zambia and the United Kingdom. The results are listed in the following section.

LIST OF SALMONELLA ISOLATED FROM WILD ANIMALS

FOREIGN

Species	*Salmonella serotype*
Bat	*S. dar-es-salaam**
Baboon	*S. isangi, S. sundsvall*
Black Fish	*S. oranienburg*
Dove	*S. typhimurium*
Echidna	*S. birkenhead*
Elephant	*S. enteritidis* var. *jena, S. muenchen, S. plymouth*
Fly	*S. anatum, S. brazil, S. decatur, S. muenchen, S. paratyphi B*
Falcon	*S. onderstepoort*
Giraffe	*S. typhimurium*
Hippopotamus	*S. aberdeen, S. bareilly, S. miami, S. uganda*
Hedgehog	*S. adamstua, S. bispebjerg, S. cerro, S. halle, S. kisarawe, S. miami*
Kangaroo	*S. chester, S. singapore*
Lizard	*S. abony, S. acres*, S. adelaide, S. aequatoria, S. agama, S. akuafo, S. alsterdorf*, S. anatum, S. atra*, S. baragwanath*, S. blukwa, S. boulders*, S. brandenburg, S. bunnik*, S. calvinia*, S. canastel*, S. cape*, S. caracas, S. carletonville*, S. chersina*, S. chester, S. christiansborg, S. claibornei, S. clovelly*, S. constantia*, S. cubana, S. dar-es-salaam*, S. dusseldorf, S. durbanville*, S. durban, S. eastbourne, S. elizabethville, S. epping*, S. erlangen*, S. ghana, S. glencairn*, S. gwoza, S. haarlem, S. hamburg*, S. havana, S. heidelberg, S. hooggraven*, S. hueningen*, S. hvittingfoss, S. inverness, S. isangi, S. kaapstad, S. katesgrove*, S. kingston, S. kintambo, S. kisarawe,*

* Subgenus II.

Species	*Salmonella serotype*
Lizard (continued)	

S. *kokomlemle*, S. *korlebu*, S. *legon*, S. *lindi*, S. *loma-linda*, S. *llandudno**, S. *magumeri*, S. *manhattan*, S. *meleagridis*, S. *merseyside**, S. *mgulani*, S. *minne-sota*, S. *mission*, S. *moero*, S. *muenchen*, S. *muizenberg**, S. *nachshonim**, S. *nairobi**, S. *neumuenster*—Subgenus II variant*, S. *nigeria*, S. *noordhoek**, S. *nordenham**, S. *odozi*—Subgenus II variant*, S. *ona*, S. *onarimon*, S. *oranienburg*, S. *orientalis*, S. *ouakam*, S. *poona*, S. *pretoria*, S. *reading*, S. *richmond*, S. *roggeveld**, S. *rooikrantz**, S. *row-barton**, S. *rubislaw*, S. *san diego*, S. *seaforth**, S. *senegal*, S. *senftenberg*, S. *simonstown**, S. *slangkop**, S. *soutpan**, S. *stevenage**, S. *sunnydale**, S. *takoradi*, S. *tamale*, S. *tennessee*, S. *teshie*, S. *tokai**, S. *typhi-murium*, S. *uphill**, S. *virchow*, S. *wandsbek**, S. *way-cross*, S. *weltevreden*, S. *westpark**, S. *woerden**, Subgenus II—1,6,14 : k : z_6 : z_{42} (un-named)

$$1,40 : m,t : z_{42}$$
$$6,7 : a : z_6$$
$$6,7 : g,t : e,n,x(z_{42})$$
$$6,7 : k : z_6$$
$$6,7 : z_{42} : e,n,x : 1,6$$
$$6,14 : k : e,n,x$$
$$17 : k : —$$
$$18 : m,t : 1,5$$
$$40 : — : 1,7$$
$$41 : k : —$$
$$42 : — : 1,6$$
$$50 : l,w : z_{42} : e,n,x,z_{15}$$
$$56 : e,n,x : 1,7$$
$$57 : g,m,s,t : z_{42}$$
$$64 : k : e,n,x,z_{15}$$
$$64 : z_{29} : —$$

Mouse	S. *anatum*, S. *brijbhumi*, S. *enteritidis* var. *jena*, S. *hessarek*, S. *teko*, S. *typhimurium*
Monkey	S. *anatum*, S. *enteritidis* var. *jena*, S. *lishabi*, S. *typhimurium*
Mole	S. *brijbhumi*, S. *vredelust**
Pigeon	S. *anatum*, S. *orion*, S. *simsbury*, S. *typhimurium*
Parrot	S. *strasbourg*

* Subgenus II.

Species	Salmonella serotype
Quail	S. halle, S. gatuni
Rat	S. agama, S. ajiobo, S. anatum, S. bukavu, S. chester, S. enteritidis var. chaco, S. nairobi*, S. nigeria, S. okefoko, S. oranienburg, S. orion, S. oslo, S. senftenberg, S. tafelbaai*, S. typhimurium, S. virchow, S. wandsbek*, S. weltevreden, S. wynberg*
Rabbit	S. anatum, S. garba, S. hofit, S. typhimurium, S. hessarek
Ruby Throat	S. gatuni
Rhinoceros	S. typhimurium
Snake	S. adelaide, S. agbeni, S. alsterdorf*, S. anatum, S. anfo, S. boksburg*, S. boulders*, S. bovis morbificans, S. braenderup, S. bredeney, S. calvinia*, S. cerro, S. chersina*, S. chester, S. djemaia, S. dublin, S. durban, S. durbanville*, S. eastbourne, S. enteritidis var. jena, S. epicrates, S. germiston*, S. gilbert*, S. heilbron*, S. helsinki*, S. houten**, S. ilala, S. infantis, S. irumu, S. javiana, S. johannesburg, S. kintambo, S. klapmuts*, S. kommetje*, S. lindrick*, S. luckenwalde, S. makumira*, S. manhatten, S. manombo*, S. mikawasima, S. muenchen, S. namib*, S. neasden*, S. new-haw, S. nigeria, S. nyanza, S. nyborg, S. oranienburg, S. othmarschen, S. pensacola, S. phoenix*, S. poona, S. potsdam, S. praha, S. pretoria, S. reading, S. rowbarton*, S. sunnycove, S. saint-paul, S. salford, S. senegal, S. stevenage*, S. umbilo, S. uno, S. vredelust*, S. wa, S. wandsbek*, S. wassenaar**, S. westpark, S. windermere, S. zanzibar

Subgenus II—1,(6),14: z_{42}: 1,6
(un-named) 42: m,t: e,n,x,z_{15}
53: z: z_6

Seagull	S. anatum, S. chester, S. isangi, S. johannesburg
Sloth	S. anatum, S. nigeria
Sparrow	S. nakuru, S. typhimurium
Skua	S. oranienburg
Secretary Bird	S. typhimurium
Turtle and Tortoise	S. abony, S. adamstua, S. ahuza, S. askraal*, S. bahrenfeld, S. bloemfontein*, S. bulawayo*, S. boecker, S. calvinia*, S. canastel*, S. carletonville*, S. cerro, S. charity, S. chersinia*, S. chicago, S. duivenhoks*, S. durbanville*, S. faure*, S. grabouw*, S. haarlem*, S. halle, S. hermannswerder, S. heron,

* Subgenus II.

FOREIGN (continued)

Species	Salmonella serotype
Turtle and Tortoise (continued)	S. hvittingfoss, S. konondoni, S. kisaware, S. klapmuts*, S. kommetje*, S. kottbus, S. lindrick*, S. madelia, S. mikawasima, S. mjimwema*, S. mobeni*, S. mosselbay*, S. nashua, S. neasden*, S. onderstepoort, S. oranienburg, S. oysterbeds*, S. panama, S. pomona, S. potsdam, S. roggeveld*, S. rowbarton*, S. rubislaw, S. ruzizi, S. san diego, S. slatograd*, S. typhimurium, S. uphill*, S. verity*, S. vredelust*, S. wandsbek*, S. weslaco, S. westpark* Subgenus II—16: b: z_{42} (un-named) 50: $1,z_{28}$: z_{42}
Toads and Frogs	S. anatum, S. hvittingfoss, S. hindmarsh
Ticks	S. klapmuts*, S. orientalis, S. roggeveld*, S. uphill*
Waterfowl	S. typhimurium

UNITED KINGDOM

Species	Salmonella serotype
Badger	S. agama, S. dublin
Blackbird	S. hessarek, S. typhimurium
Cockroach	S. typhi
Dunlin	S. typhimurium
Dove	S. litchfield
Fox	S. infantis
Finch	S. typhimurium
Grouse	S. amager, S. typhimurium
Greenfinches	S. typhimurium
Guillemot	S. typhimurium
Hedgehogs	S. brancaster, S. enteritidis var. jena
Hare	S. bredeney
Heron	S. litchfield, S. typhimurium
Knot	S. typhimurium
Mouse	S. anatum, S. bovis morbificans, S. dublin, S. enteritidis var. danysz, S. oranienburg, S. senftenberg, S. stanley, S. tennessee, S. thompson
Pigeon	S. typhimurium
Pheasant	S. bovis morbificans, S. typhimurium, S. worthington
Rat	S. bareilly, S. butantan, S. derby, S. dublin, S. enteritidis, S. enteritidis var. danysz, S. florida, S. give, S. heidelberg, S. kentucky, S. meleagridis, S. montevideo,

* Subgenus II.

Species	*Salmonella serotype*
Rat (continued)	
	S. newport, S. saint-paul, S. senftenberg, S. taksony, S. tennessee, S. thompson, S. typhimurium
Rook	*S. gallinarum*
Seagull	*S. brandenburg, S. duisburg, S. enteritidis* var. *jena, S. indiana, S. muenchen, S. panama, S. stanley, S. tennessee, S. typhimurium*
Sparrow	*S. hessarek, S. indiana, S. typhimurium*
Shag	*S. typhimurium*
Shrew	*S. typhimurium*

I would draw attention to the fact that the ticks from which salmonellae were isolated were taken from reptiles. One other point is that Salmonella subgenus II appears to occur more commonly in Africa than in other countries, certainly than in the United Kingdom, France, Germany and Australia.

HOST SPECIFIC SALMONELLAE

Certain salmonella serotypes are host specific in that they are isolated from one animal species only; rarely they may be found in a different species but such a finding usually indicates close contact with the specific host or its contaminated environment. Very often such host specific types are pathogenic to the particular host but relatively non-pathogenic to other animals. *Salmonella enteritidis* is subdivided into four varieties of which *S. enteritidis* var. *danysz* is isolated almost exclusively from wild rats and occasionally from mice. This organism has been found in outbreaks in these rodents and the death rate may be quite high. At one time baits infected with this organism were used in rat control but occasionally the infection passed to domestic animals and man so that this method of rodent control is no longer used. This relation between rats, mice and *S. enteritidis* var. *danysz* has been recorded in many parts of the world.

Wild pigeons may be infected with salmonellae but the most common infection is due to a particular variety of *S. typhimurium*, the antigenic structure being 4, 12: i: 1, 2. Farrant, Phillips and Rogers (1964) studied 246 wild pigeons trapped or shot in London and isolated this particular type of *S. typhimurium* from 43 (17%). It is generally accepted that this serotype of *S. typhimurium* is associated with all types of pigeons, including wild. This serotype has been isolated from other animal species but is rare. The common type of *S. typhimurium* has the antigenic structure 1,4,5,12 : i : 1,2, and is the most widely distributed salmonella throughout the animal kingdom as well as being the most frequently isolated type.

S. pullorum and *S. gallinarum* are well known as the cause of disease in poultry; they have also been isolated from wild birds such as wild duck, pheasant, quail, magpie, sparrow, bullfinch and curlew but not from other animal species. These findings show that these salmonellae are of importance to both wild and domestic birds. The last type I shall quote is *S. dublin*, an important cause of disease in bovine animals, which is rarely isolated from wild animals, the only ones listed being fox and rat, suggesting that these had been on farm premises.

UBIQUITOUS SALMONELLAE

There are many references to salmonella in wild animals of various species, too many to be covered in this report; a review is given by Buxton (1957) in which he lists the salmonella serotypes and the animals from which they were isolated. In the series he quotes, one finds that *S. typhimurium* is the most common infection in wild animals, being found in many varieties of birds, fox, rat, mouse, snake, cockroach and fly. Buxton also records the isolation of *S. paratyphi B* from 4 varieties of birds and a cockroach and *S. derby* from a pheasant, camel, fawn and mouse. Attention is drawn to these three salmonellae to show that they are ubiquitous, being found in many different groups of animals as are the majority of salmonellae. Data from my laboratory, shown earlier, indicate the many animals found to be infected and the variety of salmonella serotypes which have been isolated from them both in the United Kingdom and other countries.

SALMONELLA ISOLATIONS

Birds

The investigation of salmonellosis in wild animals has often been prompted by the desire to trace the source and means of spread of infection in man or in domestic animals; it is only during recent years that wild animals have been studied in their own right. The majority of such studies have been carried out on birds.

Table I shows a few of the findings in various parts of the world and indicates that *S. typhimurium*, the most ubiquitous and common

TABLE I

Birds

Ref.	Type	No. investigated	No. positive	Country	Type
Nielson (1960)	seagulls	249	4	Denmark	*S. typhimurium*
Belding (1955)	pheasants	65	5	U.S.A.	*S. pullorum*
Harbourne (1955)	rooks	79	7	England	*S. gallinarum*
Harbourne (1955)	pigeons	9	1	England	*S. gallinarum*
Harbourne (1955)	partridge	1	1	England	*S. gallinarum*
Wilson and Macdonald (1967)	birds	2715	17	U.K.	⎰ *S. typhimurium* 15 ⎱ *S. london* 1 *S. paratyphi B* 1
Goodchild and Tucker (1968)	birds	382	11	England	⎰ *S. typhimurium* 9 ⎱ *S. menston* 2

of all salmonellae, is common in wild birds. In the last two studies in Table I very many different birds were investigated and it is important to note that isolations, mainly *S. typhimurium*, were made from lesser redpoll, house and tree sparrows, lesser whitethroat, dunnock, ferral pigeons, greenfinches, tawny owl, hooded crow, rook, red-throated diver, mallard, mute swan, red grouse, herring gull, wood pigeon and gannet. I also received cultures of *S. typhimurium* from house sparrows in Bermuda. I am indebted to Mr. L. W. Cornelius (1968, personal communication) who has kindly allowed me to give some of his findings. He sent dead birds to the Lasswade Laboratory where salmonellae were isolated and identified and some of the isolations are included in the report by Wilson and MacDonald (1967). He states that an unusual number of dead wild birds, mostly greenfinches and house sparrows, were noted in Sanderstead, Surrey, in the winter months of 1966 and 1967; some of these were investigated and *S. typhimurium* was isolated. Ringing was carried out in the area and the greatest number of green-finches seen at one time in a garden was about 30. In a larger garden where ringing was not done and larger quantities of food were provided, over 200 greenfinches were counted feeding together; it was here that most of the sick and dead birds were observed. Ringing of house sparrows had been done in the area showing that house sparrows did not travel far but it was known that the resident population was about 400. In the winter months both greenfinches and house sparrows feed at the bird tables but with spring and a more plentiful food supply the greenfinches return to the woods and dead birds would not be noticed. In addition to these findings I have cultures of *S. typhimurium* from greenfinches in Melton Mowbray and in Camberley. In all these records sick and dead birds have been noted and it is reasonable to assume that salmonellosis may be an important cause of death in these birds.

Studying the varieties of birds from which salmonella have been more commonly isolated, one is struck by the fact that they are varieties which live in flocks, a habit which influences the probability of infection passing from bird to bird. I have recently been told of *S. typhimurium* infection among long-tailed tits which do not usually collect in flocks but I understand that towards the end of winter they tend to form groups and it is then that dead birds have been picked up and salmonella isolated from them. Table I shows that *S. typhimurium* was isolated from sea-gulls in Denmark and Wilson and MacDonald (1967) obtained similar results in their study.

Mammals

Some rodents are very liable to salmonella infection. It has already been stated that mice, and particularly rats, are susceptible to infection with the host specific type *S. enteritidis* var. *danysz* which causes outbreaks of disease and death in these animals. They not uncommonly pick up other salmonella types but this is related to their environment, particularly their source of food. Ludlum (1954) made a study of rats in a butcher's by-products factory and compared the findings with those isolations from rats in other premises. His studies were conducted over a period of $4\frac{1}{2}$ years in which 22% of 173 rats from the factory were infected, whereas only $4\cdot4\%$ from other premises had salmonellae. From the factory rats 11 salmonella types were isolated including *S. enteritidis* var. *danysz*, whereas from the rats from other premises two types only were found. During three months of this study the salmonella incidence rate in factory rats became very high, 40%, and this coincided with the overloading of the factory with offal and the probable extended time during which salmonella could multiply. Table II shows the number of rats infected, the variety of salmonella types isolated and also illustrates the effect of food on the infection rate in rats collected on infected premises compared with those on clean premises.

TABLE II

Salmonella in rats in the Nottingham area, England (Ludlam, 1954)

Period	Butcher's by-products factory			Other premises		
	Total rats examined	Number (and percentage) positive for Salmonella	Strains isolated	Total rats examined	Number (and percentage) positive for Salmonella	Strains isolated
Oct. 1953–Dec. 1953	60	24 (40)	11 S. enteritidis var. danysz 8 S. typhimurium 3 S. bovis-morbificans 3 S. derby 2 S. meleagridis 1 S. enteritidis var. jena 1 S. dublin 1 S. kentucky	35	1 (2·9)	1 S. enteritidis var. danysz

It is interesting to compare these results with those of Brown and Parker (1957) who found that 4·4% of 500 brown rats but none of 279 black rats were infected. Similar studies were made in Brisbane by Lee (1955) who found an infection rate of 3% and in Ibadan by Collard, Sen and Montefiore (1957) who also reviewed previous records. Lins (1964) in her study of wild animals in the Amazon forest found an infection rate of 0·6%.

Brown and Parker (1957) examined 73 mice caught on ships and isolated *S. stanley* and *S. typhimurium* from two but failed to isolate salmonellae from 168 mice in other parts of Manchester. Derrick and Pope (1960) investigated 41 wild mice caught on the Darling Downs, Australia, and from three isolated *S. bovis morbificans*. The name "*S. typhimurium*" suggests that this organism is common and causes disease in mice and experimental studies support this view; nevertheless I have been unable to confirm this from published work on wild mice and my own experience would suggest that infection in mice is comparatively rare.

Another common rodent, the rabbit, rarely becomes infected; in fact it has been said that infection only occurs when the animal is under stress, for example, when pregnant. Hares are rarely infected.

Salmonellae have been isolated from many other mammals such as the fox and hedgehog in the United Kingdom and the beaver, elephant, rhinoceros and kangaroo in other countries, to mention but a few. The fox is commonly found in farming areas where infected farm animals may be found and it is reasonable to believe that poultry could be the source of infection. The hedgehog is common in rural and even urban areas; its diet is mainly worms, slugs, snails and tiny animals. Also its proximity to the ground makes it possible for infection to occur from contaminated pasture.

Guilbride, Coyle, McAnulty, Barber and Lomax (1962) investigated 149 hippopotamus which died in the Queen Elizabeth National Park, Uganda. From 10 animals they isolated salmonellae of the following varieties: *S. typhimurium* 1, *S. aberdeen* 1, *S. miami* 1, *S. uganda* 5, *S. bareilly* 2; they also isolated *S. typhimurium* from a sick animal. These results would suggest that the animals had been in an infected environment.

Carpenter (1968) quotes surveys of wild primates from Tanganyika, Kenya, East Africa, Malaya and India; no salmonellas were isolated from 827 animals examined. On the other hand, salmonellae are not uncommonly isolated from these animals on arrival in this country, presumably after close contact with man in the collecting areas and during travel.

Reptiles

Boycott, Taylor and Douglas (1953) drew attention to the fact that tortoises, *Testudo graeco*, kept as domestic pets in the United Kingdom, contained salmonellae in their bowel in large numbers. This observation led to many investigations of wild tortoises in Israel (Hirsch and Shapiro-Hirsch, 1954), Tangiers (Vincent, Neel and Le Minor, 1960), Yugoslavia (Boycott, 1962) and Bulgaria (Dimow, 1966). The salmonella types isolated have covered a wide range; also, large numbers are excreted by the healthy animals. In some studies 96% of animals have been infected. The most commonly infected species were *T. graeco* and *T. hermanni* which are coprophagic. Boycott found that other species were less commonly infected.

Mackey (1955) examined droppings from wild house lizards *Hemidactylus mabonia* and *Mabaya striata* and found a high incidence of salmonella infection. He also examined 25 lizards and from 14 obtained salmonellae from the bowel contents. Altogether 33 salmonella types were identified and the infection rate was 48%. Collard and Montefiore (1956) examined the lizard *Agama agama* in Nigeria finding 11·2% infected. Mille, Le Minor and Capponi (1958) made similar studies in South Vietnam, finding 40% infected, and Lee and Mackerras (1955) in Australia found 62% infected. Again these appeared to be healthy animals.

Wild snakes have been examined in England (J. A. Boycott, personal communication) but were found to be negative. Le Minor, Fife and Edwards (1958) examined 310 snakes in France and isolated *S. java* from 5. Sérié and Le Minor (1959) examined snakes, colubridae and viperidae from Eritrea and Libia and found that salmonellae were more common in the latter.

Kexel and Schubert (1967) examined frogs and tadpoles in Germany but failed to isolate salmonella. Bool and Kampelmacher (1958), working in Surinam, examined 27 toads and found 6 different salmonella types in 15 which they believe was due to their frequent contact with infected material.

Insects

Insects are concerned in the salmonella story but they do not appear to play an important rôle. Floyd and Hoogstraal (1956) examined 2000 ticks collected in and around Cairo from burrows and resting places of rodents and birds. They isolated salmonella from ticks collected in hedgehog burrows and heronaries.

As stated earlier, ticks from snakes were found to harbour salmonella of the same types that were found in snakes in the area.

Eskey, Prince and Fuller (1949) cultured flea faeces and found that a number of fleas were excreting *S. enteritidis*; this infection was in the mouse colony from which the fleas originated. Huang, Chang and Lien (1937) were able to isolate *S. enteritidis* from body lice removed from patients and Milner, Jellison and Smith (1957) isolated the same salmonella from lice found on seal pups, many of which had died from this infection.

I have found it difficult to find references to salmonella in flies and, although much epidemiology has been reported relating the incidence of salmonella infection in man to the fly population, I could not find laboratory results in which salmonella had been isolated from flies. Tanner and Tanner (1953) in their book refer to a paper by De la Paz who examined flies caught in food stores in Manila and isolated salmonella from 37% of the external surfaces and 43% of the internal contents. Greenberg, Varela, Bornstein and Hernandez (1963) studied flies in a Mexican slaughterhouse in which they isolated 12 salmonella types. The highest infection was 66% in *Sepsidae* feeding on offal; 44% of *Musca domestica* feeding on offal were infected whereas 41% of those feeding on manure were infected. Flies captured on offal had somewhat higher infection rates than those taken on manure; the detailed results of this study are shown in Table III here reproduced by kind permission of the authors and the Editor of the American Journal of Hygiene.

There have been a number of reports of the isolation of salmonella from cockroaches. Antonelli (1930) in Italy studied *Blatta orientalis* caught around the open latrines of infected families and was able to isolate *S. typhi* from the feet and bodies. Mackerras and Mackerras (1948) in Australia investigated cockroaches caught in hospital wards where cases of salmonella infection were occurring; they isolated *S. bovis morbificans* and *S. typhimurium* from two species, the gut having been excised under sterile conditions. Bitter and Williams (1949) examined cockroaches caught on the kitchen porch of a nurse's home and failed to isolate salmonella but those caught mainly at sewer manholes yielded three different salmonella types.

Miscellaneous

The final group of animals which deserves mention are those that live in water, including the shellfish, mussels, cockles, oysters, clams, snails, fish and others. It is obvious from the literature that it is inferred that shellfish have been infected because human outbreaks have been related to their consumption on epidemiological grounds; also the shellfish were known to come from sewage polluted water. Nevertheless it

TABLE III

Composition of fly population and incidence of salmonellae in flies
Mexico (Greenberg et al., 1963)

Species*	Attractant	Number of flies tested	% of all flies on each attractant	Number of pools positive for Salmonella	% positive pools
Musca domestica	Manure	563	19	7/17	41
	Offal	356	8	4/9	44
Lucilia sericata	Manure	0	0	—	—
	Offal	81	2	1/3	33
Gallitroga macellaria	Manure	5	< 0.1	0/1	—
	Offal	1733†	41	14/37	38
Phormia regina	Manure	0	0	—	—
	Offal	970§	23	13/27	48
Ophyra aenescens	Manure	160	5	1/5‡	20
	Offal	351	8	4/11	36
Sepsidae	Manure	733	24	2/7	29
	Offal	391	9	6/9	66
Sphaeroceridae	Manure	1578	52	1/4	25
	Offal	300	7	1/3	33

* Other species, e.g., Stomoxys calcitrans and Sarcophaga sp., were occasionally present in very small numbers and are therefore not included in this tabulation. Sepsidae consisted of: Palaeosepsis sp., Sepsis punctum (Fabr.) and S. biflexuosa Strobl. Sphaeroceridae consisted of: Leptocera (Coproica) ferruginata (Stenk.) and L. (C.) vagans (Hal.).
† Includes one pool of 500 flies negative for salmonellae.
‡ Includes one pool of 10 flies negative for salmonellae.
§ Includes one pool of 100 flies positive for salmonellae.

has been very difficult to obtain proof of salmonella having been isolated; Macdonald, Sivell, Emms and Douglas (1948) obtained *S. brancaster, S. newport* and *S. typhimurium* from mussels taken from the sea on the east coast of England and suggested that sewage pollution was a factor. I was unable to find that *S. typhi* had been isolated from oysters although Slanetz, Bartley and Stanley (1968) record the isolation of *S. typhimurium, S. blockley, S. thompson, S. tennessee* and *S. heidelberg.* An interesting paper on this subject by Colwell and Liston (1960) shows that the external water environment does not exert absolute control over the bacterial flora of the oyster. I have the impression that the salmonella infection of oysters is a mischance which is related to the infection of their environment.

Floyd and Jones (1954) examined Nile fish, bought in the market, from which they removed the gut under sterile conditions and sought for salmonella; 13 strains were isolated of which the human specific type *S. typhi* was the most common. One must conclude that the fish were infected by the sewage contaminated water as *S. typhi* is found solely in man or as the result of human contamination.

FACTORS INFLUENCING SALMONELLA INFECTION

I have referred to the many wild animals from which salmonella have been isolated but, in order to discuss factors which have a bearing on this situation, one must refer to wild animals from which, in various studies, salmonella have not been isolated. Dr. J. A. Boycott (personal communication) made a study of small mammals and failed to isolate salmonella from any except one hedgehog from which he obtained *S. brancaster* (see Table IV). Lee and Mackerras (1955) and Goodchild

TABLE IV

Small mammals, England (J. A. Boycott, personal communication)

Species	No. tested	No. positive
Apodemus sylvaticus (long-tailed field mouse)	37	0
Clethrionomys glareolus (bank vole)	14	0
Microtus agrestis (short-tailed vole)	24	0
Arvicola amphibius (water rat)	2	0
Sorex araneus (common shrew)	15	0
Sciurus carolinensis (grey squirrel)	5	0
Talpa europaea (mole)	7	0
Erinaceus europaeus (hedgehog)	23	1
		(*S. brancaster*)

and Tucker (1968) give similar lists of various animals so that one must attempt to survey factors which influence salmonella infection in wild animals.

Adverse conditions

Reference has already been made to certain salmonella types such as *S. enteritidis* var. *danysz*, the organism common in rats and to a certain extent in mice, which are host specific. Such organisms can be found in the healthy host in non-epidemic periods and one must presume that adverse circumstances, leading to a depressed host, precipitate disease in one or a few animals which is then followed by an outbreak. Steiniger (1967) noted a high mortality among sea birds in 1955 in which he considered that although the main cause was the cold, rainy weather, the high infection with salmonella was an important contributory factor. The contamination of food, water and environment of the diseased animal is very much greater than that caused by the healthy carrier. Precipitating factors include sudden changes in temperature, lack of food, breeding and others.

The outbreaks of *S. typhimurium* infection in greenfinches and house sparrows in this country have been noted towards the end of winter when there is a scarcity of suitable food.

Feeding habits

Feeding habits influence the chances of infection with the salmonella types including *S. typhimurium* and *S. enteritidis* which have been isolated from so many different animal species. These types and others which are commonly found in man and domestic animals are transmitted from and to the wild animal. Certain sea birds choose to feed around sewage outfalls and Steiniger and Hahn (1953) isolated salmonella from sea birds probably feeding on crude sewage. The same was true of flies and cockroaches. Perhaps the best example is the study of rats by Ludlum (1954) (Table II in his paper) which shows the difference in infection rate between rats feeding on offal and others. The infection of fleas and lice appears to be directly related to infection of the animals on which they are feeding.

Environment

It is not easy to make a clear cut differentiation between food and environment but possibly the infection of fish and shellfish may be an example of this. Those that have been found to be infected may well be infected because they are surrounded by polluted water rather than the fact that they eat sewage. I know little about the food of fish so am in some doubt. It has been shown that cross infection in laboratory guinea-pigs occurs via the eye and among animals such as the hedgehog, one should consider whether in such an animal infected dust may not be a factor. The effect on salmonella infection in animals living in areas subjected to recurring floods needs consideration. Luttmann (1967) discusses this problem and the finding of *S. manchester* in local birds but not in migratory birds.

Natural behaviour

The natural behaviour of animals certainly influences the possibility of salmonella infection. Greenfinches and house sparrows appear to suffer from outbreaks of salmonellosis in the early part of the year when they flock and feed together so providing ideal conditions for the spread of infection; rats and mice also live in close proximity in groups. I do not know enough about the natural behaviour of many species discussed, but the majority of birds found negative for salmonella by Goodchild and Tucker (1968) and the small mammals by J. A. Boycott (personal communication) tend to live in greater isolation. Probably this is not the only factor for it does not account for the almost complete immunity of the rabbit. Since rabbits almost certainly meet infection by eating infected grass on farms and in areas subjected to flooding, one must presume that some animals are comparatively resistant to infection.

NATURAL GUT FLORA

Certain cold-blooded animals such as snakes, lizards and tortoises appear to have salmonella in their bowel as part of the natural flora; accepting Arizona as salmonella, it is unusual to fail to isolate these organisms from the majority of animals studied. I have been concerned with Dr. Brede in a study of snakes in Africa in which salmonella have been isolated from throat swabs and mouth swabs as well as from the intestinal contents and faeces. These animals appeared to be healthy.

It has been shown that salmonellae have been isolated from many species of wild animals, many of which were investigated in routine surveys but some being diseased. As a generalization, some birds, mammals and insects show symptoms of illness as the result of infection but many may be healthy in spite of the presence of infection; it is probable that the latter can act as a reservoir of salmonellosis.

ACKNOWLEDGEMENTS
I wish to thank Dr. J. A. Boycott and Mr. L. W. Cornelius for kindly allowing me to quote their unpublished data. Also I am grateful to Dr. Ludlum, the Comptroller of Her Majesty's Stationery Office, Dr. Varela and the Editor of the American Journal of Epidemiology for allowing me to use tables from their publications.

REFERENCES
Antonelli, G. (1930). La blatta nella igiene domestica, *G. Soc. ital. Ig.* **52**, 132–142.
Belding, R. C. (1955). The incidence of *Salmonella pullorum* in wild pheasants in Southern Michigan. *Poult. Sci.* **34**, 1441–1444.
Bitter, R. S. and Williams, O. B. (1949). Enteric organisms from the American cockroach. *J. infect. Dis.* **85**, 87–90.
Bool, P. H. and Kampelmacher, E. H. (1958). Some data on the occurrence of salmonella in animals in Surinam. *Antonie van Leeuwenhoek* **24**, 76–80.
Boycott, J. A. (1962). Salmonella species in turtles. *Science, N.Y.* **137**, 761–762.
Boycott, J. A., Taylor, J. and Douglas, S. R. (1953). Salmonella in Tortoises. *J. Path. Bact.* **65**, 401–411.
Brown, C. M. and Parker, M. T. (1957). Salmonella infections in rodents in Manchester. *Lancet* (**ii**) 1277–1279.
Buxton, A. (1957). Salmonellosis in Animals. Review Series No. 5, Commonwealth Agricultural Bureaux, England.
Carpenter, K. P. (1968). In *Some diseases of animals communicable to man in Britain*, Proceedings of a symposium, London, 1966, O. Graham-Jones (ed.), p. 35, Pergamon Press, Oxford.
Collard, P. and Montefiore, D. (1956). *Agama agama* as a reservoir of salmonella infection in Ibadan. *W. Afr. med. J.* **5**, 154–156.
Collard, P., Sen, R. and Montefiore, D. (1957). Isolation of salmonellae from rats in Ibadan. *W. Afr. med. J.* **6**, 113–116.
Colwell, R. R. and Liston, J. (1960). Microbiology of shellfish. Bacteriological study of the natural flora of Pacific oysters. *Appl. Microbiol.* **8**, 104–109.
Derrick, E. H. and Pope, J. H. (1960). Murine typhus, mice, rats and fleas on the Darling Downs. *Med. J. Aust.* **II**, 924–928.
Dimow, I. (1966). Über den charakter der fäkalen Arizona Dauerausscheidungen bei den Landschildkörten *Testudo graeco* und *Testudo hermanii*. *Z. f. med. microbiol. immunol.* **152**, 204–210.
Eskey, C. R., Prince, F. M. and Fuller, F. B. (1949). Transmission of *Salmonella enteritidis* by the rat fleas Xenopsylla cheopis and Nosopsyllus fasciatus. *Publ. Hth Rep.* **64**, 933–941.
Farrant, W. N., Phillips, A. G. and Rogers, S. M. (1964). *Salmonella typhimurium* in London pigeons. *Mon. Bull. Minist. Hlth* **23**, 231–232.
Floyd, T. M. and Hoogstraal, H. (1956). Isolation of salmonella from ticks in Egypt. *Am. J. trop. Med. Hyg.* **5**, 388–389.
Floyd, T. M. and Jones, G. B. (1954). Isolation of Shigella and Salmonella organisms from Nile fish. *Am. J. trop. Med. Hyg.* **3**, 475–480.
Goodchild, W. M. and Tucker, J. F. (1968). Salmonellae in British wild birds and their transfer to domestic fowl. *Br. vet. J.* **124**, 95–101.
Greenberg, B., Varela, G., Bornstein, A. and Hernandez, H. (1963). Salmonella from flies in a Mexican slaughterhouse. *Am. J. Hyg.* **77**, 177–183.

Guilbride, P. D. L., Coyle, T. J., McAnulty, E. G., Barber, L. and Lomax, G. D. (1962). Some pathogenic agents found in Hippopotamus in Uganda. *J. comp. Path.* **72**, 137–141.

Harbourne, J. F. (1955). The isolation of *Salmonella gallinarum* in wild birds. *J. comp. Path.* **65**, 250–254.

Hirsch, W. and Shapiro-Hirsch, R. (1954). Salmonella in tortoises. *Harefuah* **46**, 237–238.

Huang, C. H., Chang, H. C. and Lien, V. T. (1937). Salmonella infection, A study of 17 cases of *S. enteritidis* septicaemia. *Chin. med. J.* **52**, 345–366.

Kauffmann, F. (1966). The Bacteriology of Enterobacteriaceae, Munksgaard, Copenhagen.

Kexel, G. and Schubert, R. H. W. (1967). Die Darmflora von Froschen in ihrer Abhangigkeit vom Standort. *Arch. Hyg. Bakt.* **151**, 436–445.

Lee, P. E. (1955). Salmonella infections in urban rats in Brisbane, Queensland. *Aust. J. exp. Biol. med. Sci.* **33**, 113–116.

Lee, P. E. and Mackerras, I. M. (1955). Salmonella infections of Australian native animals. *Aust. J. exp. Biol. med. Sci.* **33**, 117–125.

Le Minor, L., Fife, M. A. and Edwards, P. R. (1958). Recherches sur les salmonella et arizona hebergees par les vipères de France. *Annls Inst. Pasteur, Paris* **95**, 326–333.

Lins, Z. C. (1964). Ocorrencia do Genero Salmonella em animais silvestres capturados na floresta Anazonica, *Annali. Microbiol.* **12**, 19–29.

Ludlum, G. B. (1954). Salmonella in rats with special reference to findings in a butcher's by-products factory. *Mon. Bull. Minist. Hlth* **13**, 196–202.

Luttmann, U. (1967). Untersuchungen zur Freilandbiologie der Salmonellen im Weser-Leine-Gibiet 1964–67, *Gesundhwes. Desinfekt.* **59**, Nos. 10, 11 and 12, 139–145; 159–163; 180–186.

Macdonald, A., Sivell, C. W., Emms, W. R. and Douglas, S. H. (1948). A new type of salmonella isolated from shell-fish: *Salm. brancaster. Mon. Bull. Minist. Hlth* **7**, 158–159.

Mackerras, M. J. and Mackerras, I. M. (1948). Salmonella infections in Australian cockroaches. *Aust. J. Sci.* **10**, 115.

Mackey, J. P. (1955). Salmonellosis in Dar-es-Salaam. *E. Afr. med. J.* **32**, 1–6.

Mille, R., Le Minor, L. and Capponi, M. (1958). Nouvelle contribution a l'étude des salmonella du centre et du Sud Viet-Nam. Recherches chez les lézards. *Bull. Soc. Path. exot.* **51**, 198–203.

Milner, K. C., Jellison, U. L. and Smith, B. (1957). The role of lice in transmission of salmonella. *J. infect. Dis.* **101**, 181–192.

Nielsen, B. B. (1960). *Salmonella typhimurium* carriers in Sea-gulls and Mallards as a possible source of infection to domestic animals. *Nord. Vet. Med.* **12**, 417–424.

Report, Enterobacteriaceae Subcommittee (1958). *Int. Bull. bact. Nomencl. Taxon.* **8**, 28–30.

Sérié, Ch. and Le Minor, L. (1959). Salmonella et Arizona isolées par coproculture de colubrides et Vipéridés. *Bull. Soc. Path. exot.* **52**, 133–137.

Slanetz, L. W., Bartley, C. H. and Stanley, K. W. (1968). Coliforms, fecal streptococci and salmonella in seawater and shellfish. *Hlth Lab. Sci.* **5**, 66–78.

Steiniger, F. (1967). Salmonellen bei einer Seevogel-Epizoótic auf Scharhorn. *Arch. Hyg. Bakt.* **151**, 265–278.

Steiniger, F. and Hahn, E. (1953). Über den nachweis von keimen der typhus-paratyphus-enteritis-gruppe aus vogelkot der Stord Karlsö, Sweden. *Acta path. microbiol. scand.* **33**, 401–408.

Tanner, F. W. and Tanner, L. P. (1953). Food-borne infections and intoxications. 2nd edit. The Garrard Press.

Vincent, J., Neel, R. and Le Minor, L. (1960). Les Salmonella des Tortues. *Archs. Inst. Pasteur Tunis* **37**, 187–194.

Wilson, J. E. and Macdonald, J. W. (1967). Salmonella infection in wild birds. *Br. vet. J.* **123**, 212–219.

Symp. zool. Soc. Lond. (1968) No. 24, 75–98.

LEPTOSPIROSIS IN BRITISH WILD MAMMALS

G. I. TWIGG, C. M. CUERDEN and D. M. HUGHES

Department of Zoology, Royal Holloway College,
(University of London), Englefield Green, Surrey, England

SYNOPSIS

Past or present infections with leptospira have been demonstrated in 16 out of 20 wild mammal species from mainland Britain and some off-shore islands. Eight species on certain islands have, with one exception, been shown to be free from leptospirosis. Infections have been recorded in rodents, lagomorphs, insectivores and carnivores.

Detailed studies on the Long-tailed field mouse, *Apodemus sylvaticus*, the Bank vole, *Clethrionomys glareolus*, and the Short-tailed vole, *Microtus agrestis*, have revealed inter-specific differences in infection rates and in the ability to excrete leptospires throughout life. These differences are probably related to the effectiveness of some species as maintenance hosts. Differences in infection rates between the sexes in voles, but not in mice, indicate some unknown behavioural factors. Infection rates of mice and voles increase markedly with increasing habitat wetness in some, but not all, areas. Soil and water from study areas containing infected animals has in all cases been acid although the lowest acidity recorded was from an island on which the voles did not carry leptospirosis.

Predators have evidence of past infections with several serotypes, many of which, on serological grounds, probably occur in the nearby small mammal populations.

The reservoir of leptospirosis in wildlife is probably of considerable importance to the health and performance of domesticated animals. Serological evidence of infection with a variety of serotypes has been shown in the hare and the hedgehog and 15% of thoroughbred horses from the same area had evidence of past infections with the same serotypes.

INTRODUCTION

The causative organisms of leptospirosis are spirochaete bacteria of the genus *Leptospira*. Small mammals, especially rodents, are the chief carriers of leptospires. The bacteria form colonies in the kidney tubules of the host and are shed in the urine which may contaminate streams and pools or the working environment of certain human occupations. The leptospires can obtain access to the human system across the mucosa of the alimentary canal and through skin lesions or sodden skin. Experimental infections of domestic and laboratory animals have been achieved by the same routes and in addition some animals have been infected intra-vaginally. It is not known how wild mammals become infected although there is evidence that copulation and features of habitat may have some part to play in the process.

There is some information on the course of infection in domestic animals but nothing is known about this matter in wild animals. In man, after a period of incubation varying from 4 to 19 days, the onset of illness is often sudden. Jaundice appears towards the end of the first week and increases during the second week which is the critical period. The most serious risk to life is renal failure. Leptospires are found in the blood during the first week and become progressively less later while in the urine they are unusual in the first and second weeks and become more abundant later. Antibodies to *Leptospira* are found in the patient's serum from about the seventh day of illness in most cases. In those domestic animals which have been studied the course of illness broadly follows this pattern.

The many types of *Leptospira* cannot be differentiated on morphological grounds but they do produce different serological reactions in the host. For this reason the term serotype has been coined. There are over 60 serotypes. Some, e.g. *Leptospira icterohaemorrhagiae* carried in the Brown rat (*Rattus norvegicus*), are cosmopolitan whilst at the other end of the scale one serotype is severely limited in range.

Prior to 1958 it was believed that the only pathogenic leptospires in the British Isles were *Leptospira icterohaemorrhagiae* in *Rattus norvegicus* and *L. canicola* in dogs. Infection by the former serotype had also been found in a wild fox and in silver foxes. It was thought that the organisms did not reach Western Europe until after the separation of Britain from the Continent had taken place and hence, with the exception of the rat and the dog which came later and brought their own serotypes with them, our wild mammals did not harbour leptospires.

Elton, Ford and Baker (1931), in the course of an investigation into the health and parasites of a wild field mouse (*Apodemus sylvaticus*) population, found that 29 out of 356 field mice in the Oxford area carried leptospires. None were isolated and no serological tests were made but the authors inoculated guinea pigs with infected kidney material without conspicuous success. They observed that these mouse strains of leptospires were less pathogenic to guinea pigs than were the local rat strains.

Between that study and 1958 the only wild mammal investigated for leptospirosis in Britain was *Rattus norvegicus*. Middleton (1929), Broom and Gibson (1953) and Broom (1958) published data on infection rates in the rat. Broom and Coghlan (1958) examined 162 specimens of small wild mammals near Glasgow in Scotland. They succeeded in isolating serotypes of leptospires hitherto unknown in the British Isles. *L. ballum* was cultured from two field mice (*Apodemus sylvaticus*), one Short-tailed vole (*Microtus hirtus*) and one Bank vole (*Clethrionomys glareolus*). Six strains isolated from *Microtus* and *Clethrionomys* were related to

L. sejroe and *L. saxkoebing* and the sera of 7 out of 162 rodents contained agglutinins for *L. sejroe*. One of these sera also agglutinated *L. bataviae* and another *L. grippotyphosa*. Broom and Coghlan (1960) isolated *L. bratislava* from the hedgehog (*Erinaceus europaeus europaeus*) in Scotland. This serotype, which is closely related to *L. australis A*, is the type most commonly found in hedgehogs in Czechoslovakia. Slavin, Salt and Wolff (1964) isolated *L. erinacei auriti* from the Water vole (*Arvicola amphibius*) at Uxbridge, near London.

The wildlife studies so far in Britain have either presented information on infection rates of *Rattus norvegicus*, usually a sample taken over a short period of time, or have extended the serotype list. Prior to the study reported here, no extensive wildlife surveys have been attempted.

The importance of leptospirosis in farm animals has likewise not been fully appreciated in Britain despite the evidence of abortion and infertility due to leptospirosis in other countries. Evidence of leptospirosis in British domesticated animals has been shown by several workers. Seiler, Norval and Coghlan (1956) found antibodies for *L. canicola* in sera from 46 out of 76 pigs in one piggery and Alston and Broom (1958) demonstrated agglutinins to *L. icterohaemorrhagiae* and *L. canicola* in a large proportion of 108 healthy horses.

Michna (1958) demonstrated antibodies to *L. icterohaemorrhagiae*, *L. canicola* and *L. pomona* in slaughter pigs slaughtered in Scotland and the same author (1967) published results of a serological survey of pigs, dogs, cattle, sheep, horses and roe deer. Members of all these species had antibodies to a variety of *Leptospira*. Especially interesting were the pigs—in 601 specimens of serum obtained from sows with a history of abortion and breeding difficulties, 293 (48·7%) contained antibodies. Altogether, out of 3061 pig sera 1184 (38·6%) reacted with one or other of four serotypes.

Since antibodies to serotypes isolated from small mammals in that part of the British Isles were present in the farm animals investigated by Michna (1967), the importance of surveys to investigate the occurrence of leptospirosis in wildlife is evident.

MATERIALS AND METHODS

Almost 1700 specimens from 20 wild mammal species have been collected from mainland Britain and some islands off the Pembroke and Scottish coasts. A further 220 specimens from domestic animals have been examined. Detailed studies have been carried out on three commonly occurring rodent species in the vicinity of this laboratory in

Surrey—the Long-tailed field mouse (*Apodemus sylvaticus*), the Bank vole (*Clethrionomys glareolus*) and the Short-tailed vole (*Microtus agrestis*). Further samples of *Microtus* have been collected in Derbyshire.

Small mammals are live trapped in Longworth traps and returned to the laboratory. After killing they are weighed, sexed and measured and external parasites removed. Heart blood, kidney tissue and urine are examined by dark-field microscopy for motile leptospires. Korthof culture medium is inoculated with kidney tissue and kept at 28–32°C for 4 months and then discarded if no leptospires are seen. One half of each kidney is fixed in 10% formal saline, embedded in paraffin wax, sectioned at 5μ and stained by the Warthin-Starry silver technique for spirochaetes. Blood serum is refrigerated and tested by means of agglutination tests for antibodies to 15 serotypes. Data on vegetation, soil acidity and water content are collected in all study areas.

For many animals there are, therefore, 6 lines of evidence. For various reasons full data are lacking in some cases, usually where other workers have volunteered to collect material in the course of their own studies.

RESULTS

Wild life survey

Rodents in Surrey study area

Apodemus sylvaticus. This is one of the most common and widely distributed small mammals. It is nocturnal and is found in woodland, hedgerows and fields and will occupy buildings in the absence of the House mouse (*Mus musculus*).

The infection rate in *Apodemus* increases with body weight and hence age. The exception to this is in females in the highest weight group where infection is low although in all other weight groups it closely parallels that of males.

It is probably rare for field mice to survive two winters. Those born early in the season will breed in the same year and perhaps die off late that winter. Young reared late in the year will not breed that year but will be the main breeding stock for the next season. The overwintering weight is in the region of 15 g. The infection rates of animals below 15 g (juveniles) and above that weight (adults) have been examined separately.

Animals of less than 15 g have been caught from May onwards, in small numbers between May and August and in larger numbers from October onwards. Infected immatures have not been found before

TABLE I

Infection rates in Apodemus sylvaticus, Clethrionomys glareolus and Microtus agrestis

Species	No. animals	No. of infections diagnosed by:						Percent females infected	Percent males infected
		Culture	Dark-field	Serology	Histology	All methods No.	%		
Apodemus sylvaticus	248	38	43	31	24	64	25·8	25·3	26·6
Clethrionomys glareolus	151	11	17	8	8	30	19·9	25·9	12·9
Microtus agrestis	86	5	—	14	1	16	18·6	22·4	13·5

November and are found in November, December and January, after which time they qualify for the higher weight group.

Increasing numbers of adults are caught from June onwards until October when the population reaches a peak. Numbers then decline slowly throughout the winter. The infection rate is low from April to August and increases markedly in September, remaining high until April when it falls sharply. This is presumably because older animals die off at this time. The maintenance of infection is carried on throughout the summer months by young animals infected the previous autumn and remains low until autumn when animals born early that season become infected.

So far 63 animals have provided evidence of leptospirosis. It is seen (Table I) that almost twice as many are culturally positive as have antibodies. Antibodies to a wide variety of serotypes are demonstrated in the *Apodemus* sample. The serological data are presented in Table II.

TABLE II

Apodemus sylvaticus: *serotypes and titres with number positive*

| Serotype | Titre | | | | | |
	1:30	1:100	1:300	1:1000	1:3000	1:10 000
icterohaemorrhagiae	7	4				
bratislava	1	2	5			1
canicola	2	6	3	1		
grippotyphosa	4	3	1			
australis		1				
bataviae	1	3		1		
erinacei	2					
pomona	1					
poi		2				
autumnalis		1	1		1	
ballum		1				

It has been possible in *Apodemus* (and *Clethrionomys* and *Microtus*) to examine certain features of infection which, because of the limited data, is not possible for other species. One point of interest in *Apodemus* is the presence of antibodies in animals in which there are leptospires circulating in the blood stream. Ten mice had leptospires in the blood demonstrated by dark-field examination; 3 of these had titres of 1:300 to *L. bratislava*, 1:30 to *L. icterohaemorrhagiae* and 1:30 to *L. grippotyphosa*. It is even more interesting that isolates were made from the kidneys of two of these animals.

Although 6 voles carried leptospires in the blood, none of them had antibodies.

Clethrionomys glareolus. The Bank vole is found in the same general habitat as *Apodemus* although it is most numerous where the ground flora is dense and especially when ground ivy (*Glechoma hederacea*) is present. It forms the second largest sample for which full data are available. Infection data are presented in Table I.

There are certain features of this species in which it differs markedly from *Apodemus*. Firstly, it is noteworthy that the highest infection rate is seen in animals in the smallest weight group and the rate decreases throughout life. Nearly three times as many animals excrete leptospires as carry antibodies. In addition the infection rate between the sexes varies considerably; 26·3% of males are infected compared with only 11·6% of females.

A small number of animals possess antibodies to *L. bratislava, L. canicola, L. icterohaemorrhagiae, L. grippotyphosa* and *L. bataviae.*

Microtus agrestis. The Short-tailed vole occupies meadows and grassland; 76 animals have been examined, 23 from Surrey and 53 from an area of afforestation in the Pennines between Sheffield and Manchester. Table I presents data on the pooled sample.

Infection is highest in animals in the 20–24 g weight group, increasing with age to that point after which there is a fall off in the rate. The most interesting feature about this species is that the antibody rate far exceeds the culture rate which indicates that fleeting infections are the rule in *Microtus* but that evidence of this persists in the form of antibody. As in *Clethrionomys*, males have a higher infection rate than females.

Diagnostic titres. The interpretation of the serology of wildlife surveys presents many problems, the most important of which is to decide what is the minimum titre that can be regarded as indicative of infection. On this point there is no hard and fast rule. Various workers have used minimum titres ranging from 1:20 to 1:400; many have ignored the matter. Recently, Michna (1967) reviewed the literature on diagnostic titres in a paper on the serology involved in an extensive survey, dealing mainly with domestic animals.

Broom (1959) says "In serological surveys titres of 1:100 or more are accepted as significant by most workers and it is usual to find that the serum reacts with only a single serotype, or a few closely related ones". The majority of wildlife survey workers have however taken a titre of 1:30 as the minimum indicative of infection. This level has been applied to the data on *Apodemus* with the following results: 30 mice had titres of 1:30 and higher. Nine of these had titres of 1:30 but 6 of them were positive in other ways, either by dark-field, histology or culture;

F

had this been solely a serological survey these might have been excluded from the results. Only 3 animals had titres of less than 1:100 without other supporting evidence. Thirty-three animals were positive in various ways (culture 20; histology 2 and dark-field 10) but had no titres at all.

For *Apodemus*, then, the acceptance of a level of 1:100 would have excluded 6 which were positive with titres of 1:30. Perhaps the other 3 were carrying leptospires in small numbers but they were not isolated. There seems to be no reason, at any rate with *Apodemus*, to exclude animals with titres of 1:30. There is not enough data on other species to submit them to the same yardstick but a provisional level of 1:30 has been taken for all serological material in this survey.

Other rodents

Other species are usually in small numbers for each locality other than islands off the coast. The island fauna is dealt with separately.

House mouse (*Mus musculus*). Only 5 have been examined, all from Surrey. Two were infected. In one animal leptospires were found in kidney culture subsequent to post-mortem examination but the strain died out. In the second, colonies of leptospires were seen in kidney sections but could not be cultured. No antibodies were demonstrated.

Yellow-neck mouse (*Apodemus flavicollis*). This species occurs in small numbers in field mouse populations. Occasionally, isolated pockets of larger numbers are found but comparatively little is known of the biology of the species. Three specimens have been trapped locally, one being sero-positive with titres of 1:100 to *L. canicola* and 1:300 to *L. bataviae*. It is noteworthy that an *Apodemus sylvaticus* trapped 80 yards away had antibodies to the same serotypes.

Water vole (*Arvicola amphibius*). The Water vole inhabits slow-moving streams and rivers in the banks of which it burrows and lives. Twenty-five animals (17 males and 8 females) were trapped from Colnbrook in Buckinghamshire, near London Airport. Three animals were sero-positive, 2 having titres of 1:100 to *L. canicola* and the third with a titre of 1:300 to *L. icterohaemorrhagiae*. In a fourth animal motile leptospires were seen in the kidney with dark-field microscopy but they could not be cultured. This animal was not sero-positive.

Preserved kidneys and serum were examined from 7 voles caught at various sites in Norfolk. One animal had titres to the following serotypes: *L. icterohaemorrhagiae* (1:30), *L. autumnalis* (1:30), *L. erinacei* (1:100), *L. hardjo* (1:100) and *L. saxkoebing* (1:30). On these limited data the results (1 positive out of 7) are on the conservative side.

Grey squirrel (*Sciurus carolinensis*). This species is now present over much of Britain where it has replaced the native Red squirrel (*Sciurus vulgaris*). Fourteen specimens from Windsor have been examined; one had a titre of 1:30 to *L. canicola*.

Coypu (Myocastor coypus). Introduced into Britain for fur farming, the coypu now lives feral over large areas of East Anglia where the river systems and dense vegetation provide a suitable habitat. Six specimens, 4 from Norfolk and 2 from Suffolk, were examined; all were infected.

By dark-field microscopy motile leptospires were seen in the blood of 3 and in the kidneys of 2. Three were sero-positive and histological examination revealed colonies in the kidney tubules of 2 animals. The 3 sero-positive animals had antibodies to *L. icterohaemorrhagiae* (1:30, 1:100, 1:1000); *L. bratislava* (1:100, 1:300) and *L. erinacei* (1:100). All attempts at culture failed. It is very difficult to isolate leptospires from coypu because of the high degree of contamination of the kidneys of this species with other bacteria (L. H. Turner, personal communication).

Lagomorpha

Rabbit (Oryctolagus cuniculus). Samples have been collected from Skokholm Island, Pembroke, and from a nature reserve in Buckinghamshire. Only histological and serological data are available for both samples.

From Skokholm 57 specimens were available (24 males and 33 females). Ten of the males were infected; 4 were sero-positive, 3 with titres of 1:30 to *L. icterohaemorrhagiae* and 1 with a titre of 1:30 to *L. bataviae*. Three of the sero-positive animals were also positive histologically. A further 6 animals were histologically positive although none had antibodies.

Nine of the females were positive. Two had titres of 1:100 to *L. icterohaemorrhagiae* and both were histologically positive. A further 7 without antibodies were histologically positive.

Nineteen out of 57 (33·3%) were positive on this limited evidence. The interesting thing is the preponderance of animals with leptospire colonies present in the kidney tubules but without antibodies, although some had both.

From mainland Britain only 16 rabbits (from Slough, Bucks.) have been examined. These were negative in all respects.

Hare (Lepus europaeus). Three main collecting areas have been studied; one on the Berkshire Downs, one at Balsham in Suffolk and the third at Bartlaw in Norfolk. In all three cases the hares were caught on farmland. Serological and histological examinations are now complete for the Balsham sample but so far serum only has been examined from the other two.

From a sample of 225 hares shot on the Berkshire Downs, 42 were sero-positive. The range of antibodies is shown in Table III.

TABLE III

Hare (Lepus europaeus) *from Compton, Berkshire. Serotypes and titres with number positive*

Serotype	Titre			
	1:30	1:100	1:300	1:1000
bataviae	14	8	3	—
canicola	12	8	1	—
pomona	5	3	1	—
bratislava	1	1	3	—
icterohaemorrhagiae	—	—	1	1

From Balsham 61 specimens (5 males and 56 females) were collected of which 20 (32·7%) were infected. The males were uninfected and 18 females were sero-positive, only 2 having leptospire colonies in the kidney tubules. Neither of these was sero-positive. Table IV lists the serotypes and titres.

From Bartlaw, 7 males and 26 females were collected. Nine females were sero-positive (Table IV).

TABLE IV

Hare (Lepus europaeus) *from Balsham and Bartlaw, Nr. Newmarket. Serotype and titres with number positive*

Serotype	Titre			
	1:30	1:100	1:300	1:1000
canicola	9	12	4	—
bataviae	12	3	1	—
grippotyphosa	3	—	—	—
poi	1	—	—	—
saxkoebing	1	—	—	—

In addition to these, 8 females were collected from Ashton Wold, Oundle. None of these was infected but a female hare from Egham, Surrey, had titres of 1:30 to *L. icterohaemorrhagiae* and 1:30 to *L. pomona*.

Insectivora

Common shrew (Sorex araneus). These have been caught incidental to other trapping programmes and no effort has yet been made to trap specifically for shrews. All have been caught in Longworth traps baited for rodents and hence were usually dead when the traps were examined each morning. As a result data are fragmentary but the favourable results so far indicate that this species would repay further study. Being very small anyway, there is rarely enough serum for agglutination tests from a specimen that has been dead some hours.

Twenty-two animals from the Surrey study areas were examined, of which 5 were positive in various ways. Table V presents data on the species. An isolated specimen from Derbyshire was sero-positive with a titre of 1:30 to *L. pomona*.

TABLE V

Infection in the Common shrew (Sorex araneus) *and the Hedgehog* (Erinaceus europaeus europaeus)

	Number of animals	No. infections diagnosed by:				
		Culture	Dark-field	Serology*	Histology	All methods
Common shrew	22	4	4	1	1	5
Hedgehog	14	2	2	6	—	7

*No serum could be obtained from 14 of the shrews.

Mole (Talpa europaea). The few specimens so far studied have been caught by mole catchers and have thus been dead for some time prior to examination. One out of 6 taken on the Surrey study area was sero-positive to *L. grippotyphosa* (1:30) and none from 9 on a Surrey farm.

Hedgehog (Erinaceus europaeus europaeus). The material studied consists of isolated specimens from a variety of localities in addition to 56 serum samples from Compton, Berkshire.

Agglutinins to 5 serotypes were found in the latter animals (Table VI) and 26 animals were sero-positive. It is unfortunate that the serum had been collected for another purpose without data on weight and sexual state and so there is no way of relating infection to sex and age.

Complete data are available on a further 14 animals from a variety of localities in Southern England. The infection data of this sample are summarized in Table V.

Titres were recorded in these animals to *L. bratislava* (1:100, 1:300), *L. pomona* (1:30), *L. canicola* (1:30, 1:100) and *L. icterohaemorrhagiae* (1:100).

TABLE VI

Hedgehog (Erinaceus europaeus europaeus). *Serotypes and titres with number positive*

Serotype	Titre				
	1:30	1:100	1:300	1:1000	1:3000
bratislava	4	6	6	2	2
icterohaemorrhagiae	2	3	—	—	—
poi	1	—	1	—	—
autumnalis	1	1	—	—	—
erinacei	1	—	—	—	—

Carnivora

Three species of carnivore have been examined. In the main these animals have either been shot or have come to some other violent end. As a result serum samples have been the main material although some kidneys have been prepared histologically.

Fox (*Vulpes vulpes*). Adjacent to the Surrey study area is a large tract of land where the fox population is controlled by regular shooting. Twelve foxes have been examined and serology completed for all. Seven males (4 infected) and 5 females (all infected) comprise the sample. Serological data are given in Table VII where it will be seen that antibodies to several serotypes are found in some animals. On serology alone 9 out of 12 (75%) were positive.

A sample of 6 foxes from Orpington, Kent, was negative serologically.

Badger (*Meles meles meles*). Two specimens were examined from the same site as the sero-positive foxes. In one animal titres to *L. icterohaemorrhagiae* (1:30) and *L. erinacei* (1:30) were found. The other was sero-negative.

Two other badgers, one from Chessington, Surrey and the other from Salisbury, Wilts., were uninfected.

Weasel (*Mustela nivalis*). Three females for which full data are available have been caught at sites in Derbyshire, Surrey and Oxford. The latter specimen had a titre of 1:100 to *L. bratislava*.

TABLE VII

Titres and serotypes in foxes (Vulpes vulpes) from Windsor

Animal No.	Icterohae-morrhagiae	Serotype								
		canicola	poi	autumnalis	erinacei	ballum	hardjo	saxkoebing	hebdomadis	bataviae
744	1:30	—	—	—	—	—	—	—	—	—
841	—	1:30	—	—	—	—	—	—	—	—
843	1:30	—	1:30	1:30	1:30	—	—	—	—	—
848	—	—	1:100	1:100	—	—	—	—	—	—
742	1:50	—	—	1:500	—	—	—	—	—	1:100
741	—	—	1:30	1:30	—	—	—	—	—	—
A.119	1:30	—	—	—	—	—	—	—	—	—
A.632	1:3000	—	—	—	1:100	—	—	—	—	—
A.14	—	1:30	1:100	1:100	—	1:30	1:30	1:1000	1:30	1:100

Ecological aspects of leptospirosis

Infection rates and habitat

In Surrey woodland. In each trapping area the exact location of all animals examined is recorded in relation to vegetation, water courses and other features of the habitat. So as not to reduce the population too drastically, trapping has been restricted to four trap nights in each study sector and each sector is trapped only once every two years.

The Surrey study area consists of four woods, 2 of them (Alderhurst and College Wood) on the upper slope of the valley side and 2 on the lower slope and valley floor (Grendon Wood and Great Wood). A brief description of each follows:

College Wood. A central region of coniferous and natural deciduous trees is marshy due to ponds and streams. Strips of woodland connect this with an outer very dry region of deciduous woodland.

Alderhurst. Consists of deciduous woodland and brambles in a series of strips and copses between which is grassland.

Grendon Wood. Mainly oak and birch woodland with thick brambles and bracken. A central hill slopes down to east and west and at the bottom of each slope a stream runs due south. Beyond the west stream the land rises to the edge of the wood.

Great Wood. Mainly oak and birch with bracken and brambles as the main ground flora. Flat and dry, it is separated from Grendon Wood by a 10-acre field.

The infection rates of animals trapped from these 4 areas are as follows:

College Wood	—	30·6% infected
Alderhurst	—	14·5% infected
Grendon Wood	—	25·1% infected
Great Wood	—	30·8% infected

If data had been collected without reference to location and habitat, these figures would disguise several interesting features of the infection rates in this area.

The rate is lowest in Alderhurst and in this area infected animals are fairly evenly distributed amongst uninfected ones, at least so far as trap sites show. In the other 3 areas the distribution of infected animals is uneven. In Great Wood trapping took place in Autumn and 52 rodents were caught (26 *Apodemus* and 26 *Clethrionomys*). In one small sector which comprised approximately one-seventh of the whole area, 14 of the total number of 16 infected animals (87·5%) were caught (7 *Apodemus* and 7 *Clethrionomys*). Only 2 uninfected animals (*Clethrionomys*) were trapped in the same area and over the whole of the major part of the trapped area only 2 more infected animals were found. In terms of soil acidity, moisture and ground flora all parts of the total area are alike and at the moment it is not possible to suggest any reason for this uneven distribution. Since both species were equally infected, some factor of the environment common to both is indicated.

In Grendon Wood the habitat falls naturally into 3 parts: the main stream and west slope; the central elevated part; the east valley and minor stream. The terrain along the main stream is swampy and most of the rodents caught in that section were trapped along the stream banks and marshy grassland. Oddly enough the infection rate is very low there, only 4 out of 55 (7·3%). Further east on the dry hill the infection rate goes up to 23·5% whilst in the slightly damp east valley it is 47·9%. In view of what follows in the next section, the low infection rate in the wettest part is hard to understand.

Infection and wet habitats. Since leptospires cannot tolerate drying out, wet soil and vegetation should favour the maintenance and transmission of the organisms in the environment. There is little information in the literature on the subject of infection rates and the habitat. Ferris and Andrews (1967) found that in skunks and opossums infection with leptospires and their excretion was heavier in the colder and wetter months of the year.

In College Wood the infection rates for *Apodemus* and *Clethrionomys* trapped from the central marshy area and the outer dry area differ greatly. The water content of soil from the 2 areas has been measured by taking soil samples at a time when there had been little rain for some time previously and evaporating them to dryness. It is seen from the following that the infection rate increases as the water content of the soil increases:

Percentage infected rodents	Percentage soil water
4·5	26·3
22·2	32·6
69·2	42·3

Soil acidity. Acid soils are believed to be unfavourable to leptospires although some evidence has been put forward by Smith and Turner (1961) to show that soil pH may be of less importance in the maintenance of leptospires than is supposed.

The pH of soils, ponds and streams on all trapping areas is recorded. All soils studied so far, both of areas with high infection rates and those with low, have been acid although variation in acidity occurred. Soils on the Surrey study areas have an acidity which ranges from 5·0 to 6·0,

TABLE VIII

Island species and localities

Common name	Species	Island	Locality
Rabbit	*Oryctolagus cuniculus*	Skokholm	Off Pembroke
House mouse	*Mus musculus*	Skokholm	Off Pembroke
Skomer vole	*Clethrionomys glareolus skomerensis*	Skomer	Off Pembroke
Great black-backed gull	*Larus marinus*	Skomer	Off Pembroke
Red deer	*Cervus elaphus*	Rhum	W. Scotland
Long-tailed field mouse	*Apodemus sylvaticus fridariensis*	Rhum	W. Scotland
House mouse	*Mus musculus*	Fair Isle	N. Scotland
Long-tailed field mouse	*Apodemus sylvaticus hebridensis*	Fair Isle	N. Scotland
Long-tailed field mouse	*Apodemus sylvaticus fridariensis*	Fair Isle	N. Scotland

whilst in the Derbyshire uplands the pH was 4·2 to 4·8 and on Skomer Island 3·6 to 5·0. In Derbyshire the *Microtus agrestis* had an infection rate of 21% although the species clearly does not function as a maintenance host. The most acid soils encountered have been those with a pH of 3·6 on Skomer.

Island faunas

Nine species from four offshore islands have been studied. Table VIII lists the species and localities.

The only island mammal to show any evidence of leptospirosis is the rabbit on Skokholm (*vide supra*). Full data are not available for some species but there is positive evidence for some mainland species based on smaller samples. Out of 29 specimens of the Great Black-backed gull (*Larus marinus*) on Skomer two specimens had titres of 1:100 to *L. canicola*.

Wildlife, domestic animals and man

The first stages of a survey on the Berkshire Downs into leptospirosis in a cross section of animal life has been completed. Species as varied as the hedgehog, hare, thoroughbred horse and man have been investigated. This area is one of the most important centres of race horse training in the country.

The survey was prompted by the initial examination of serum from a sick horse with biochemical evidence of liver damage and a recent poor performance when racing. The animal had a positive titre of 1:3000 with both *L. bratislava* and *L. pomona* antigens. Prompt treatment with antibiotics was successful and in the 7 months subsequent to this the titre to *L. bratislava* fell to 1:300 whilst the *L. pomona* titre had disappeared after 6 weeks.

Since then, serum has been taken from horses in a number of stables in the area. The results are of great interest since the antibodies demonstrated in these horses followed very closely the pattern already seen in 226 hares and 56 hedgehogs collected from the same area. Recently serum has been collected from a small number of stable attendants and several small mammals in the area. The small mammal sample has proved, surprisingly, negative. The hedgehogs (*vide supra*) had agglutinins to *L. bratislava* in 20 out of 26 sero-positive animals. Out of 42 infected hares, 5 had titres to *L. bratislava*, 9 to *L. pomona* and the majority were sero-positive to *L. bataviae* and *L. canicola*. Of the 211 horses from the area, 7 had titres to *L. bratislava* (8 if the original sick animal is included) and 15 to *L. pomona*.

One stable in particular had 9 reactors to *L. pomona* out of a total of

59 horses whilst in addition a titre of 1:100 to the same serotype was demonstrated in one of the 13 stable attendants.

Titres to *L. canicola* were shown in 21 of the hares from the area whilst 11 horses had positive titres to this serotype.

Titres to *L. icterohaemorrhagiae* were seen in a few horses as well as in a small number of hedgehogs and in 2 hares.

DISCUSSION

This survey has provided some evidence of the extent of leptospirosis in wild animals from different areas in the British Isles. It is perhaps important in trying to assess the rôle of wildlife in leptospirosis to distinguish between those species functioning as maintenance hosts and those which are only incidental hosts. Effective maintenance hosts carry and excrete the organisms throughout most of their lives, e.g. *Rattus norvegicus*, whilst incidental hosts excrete leptospires for a short period only. It is believed that voles and mice do not excrete their leptospires for so long as rats (relative to their life span) although it is known that *L. ballum* may be excreted for over a year (in van der Hoeden, 1958). Borg-Petersen (1944) recorded lifelong excretion of *L. saxkoebing* by mice and Ananin (1954) observed leptospiruria in *Microtus* during a period of 7 to 377 days, with an average of 1 to 3 months.

In the present study it has been seen that in *Microtus*, where antibody rates exceed culture rates by almost 3 to 1, infections are probably of short duration and hence self-curing. This species is therefore not an important maintenance host. In *Clethrionomys* the position is quite different. In this species the culture rate is high and the serology rate low and in this case $2\frac{1}{2}$ times as many animals excrete leptospires as carry antibodies. In *Clethrionomys* the infection rate is high in young animals and falls later but, although fewer old animals are infected, leptospires are excreted in all age groups. It would appear, however, that self-curing of infections does, on occasion, take place and also that antibody does not persist.

Culture and serology rates are more nearly equal in *Apodemus* but more animals are excreting leptospires than carrying antibodies and excretors occur in greater numbers in each weight group to old age. *Apodemus* therefore appears to be a lifelong excretor.

The high infection rate in the young *Clethrionomys* indicates that some feature of the biology of this species in relation to the acquisition of leptospirosis is quite different from either *Apodemus* or *Microtus*. It is tempting to speculate that infection of the young may perhaps be acquired from the parents whilst still in the nest.

Unknown factors in the habits of *Clethrionomys* and *Microtus* might be responsible for the higher infection rates of males than females. A similar situation occurs in leptospirosis infection of dogs (van der Hoeden, 1958). That author pointed out that infections (by *L. canicola*) from dog to dog are strongly promoted by the sexual rites customary in dogs. The male uses his nose and tongue for exploration of the female genitalia and thus comes into contact with freshly voided excretions. Differences in the sexual behaviour of mice and voles might likewise account for observed differences in infection rates in this study.

It is also known that male mammals tend to range further than females and hence would have more opportunity of becoming infected but this would not explain the lack of differences between male and female *Apodemus*.

There is surprisingly little information on the method of intra-specific transmission of leptospires. It has been suggested (Ostertag, 1950) that the route is venereal since in *Rattus norvegicus* infection begins at sexual maturity. This would appear unlikely in *Apodemus*. Animals of less than 15 g (those born late in the season) become infected in the Autumn of the same year. These animals will not be sexually mature until the following Spring. Furthermore, in animals heavier than 15 g which have been breeding in the Summer, the infection rate is low from April to August and begins to increase from September onwards. This would also indicate some unknown seasonal factor in infection. It is well known that in Northern Europe epidemics of leptospirosis in man and domestic animals occur in the Autumn when populations of rodents are at their peak. Whether this is due to some feature of the environment or to the fact that with increased density of animals the chances of contacting infected material are greater is not known.

In Wisconsin, Trainer, Hanson, Pope and Carbrey (1963) report that in White-tailed deer (*Odocoileus virginianus*) there is greater infection of young females than of young males. This is apparently due to the fact that young females can and do breed during their first rutting season, while male fawns do not. Experimental infections indicated that deer were susceptible to *L. pomona* by the intravaginal route (Trainer, 1961).

Ananin (1954) attached great importance to sexual contact as a main factor in the spread of leptospirosis among *Microtus oeconomus*. The highest incidence occurred during the summer months (the main reproductive season) in sexually mature animals. Doeleman (1932) reported a case of a woman who acquired leptospirosis 10 days after copulation with her husband who was convalescent from the disease and was still in the stage of leptospiruria.

The data on the spatial distribution of infected animals in relation to the habitat in College Wood indicate that a wet habitat may be important in infection. In several countries a highly significant correlation has been noticed between rainfall and the appearance of leptospirosis some 10–12 days later (in van der Hoeden, 1958). It is interesting that surveys on the infection rates of rodents with leptospires reveal very low rates in the dry parts of the world. In the Sudan no *Rattus norvegicus* examined carried leptospires (Kirk, 1938). In Egypt 0·5% of the same species, 1·0% of *Rattus rattus* and 0·1% of *Arvicanthis niloticus* carried *L. icterohaemorrhagiae* (McGuire and Myers, 1957). Of course, rodent species vary from place to place but the Brown rat is cosmopolitan and detailed surveys of leptospirosis infection in this species from widely differing climatic zones would be very useful.

Although an increased infection rate with increasing habitat wetness has been observed in College Wood, the lowest rate in Grendon Wood was seen in the wettest part. Furthermore, the uneven distribution of infected animals in Great Wood did not appear to be related to soil water or indeed to any of the parameters that have been measured. The fact that two species were involved in the latter example would indicate that some factor in the environment common to both of them was involved.

The relationship of infection rate to soil acidity is by no means clear. Nearly all the soils so far tested have been acid, often strongly so, yet leptospirosis has been present in the local mammals. Soil acidity is a complex subject. Smith and Turner (1961) studied the survival of four leptospire serotypes from Malayan rodents in buffered distilled water at pH's ranging from 5·3 to 8·0. All survived longer in alkaline than in acid water. When they studied survival in aqueous extracts of soil samples from different areas in Malaya no correlation was found between pH and survival time. They found that bentonite clay adsorbed about half the leptospires in suspension. It is known that the pH of the water "shell" surrounding small particles is different from that of the general body of soil water, the hydrogen ion concentration (acidity) increasing the nearer to the particle one measures. Adsorption would therefore place leptospires in a more acid environment.

The predator–prey relationship has received little attention in leptospirosis studies although the main reservoir hosts of leptospires, the rodents, form the major part of the diet of small carnivores. Foxes from land adjacent to the Surrey study area have shown serological evidence of past infections with several serotypes, one animal in particular having titres to no fewer than 8, and these with a few exceptions are revealed by serology in the local small mammals. The presence of antibodies to *L. icterohaemorrhagiae* in 5 of the 12 animals indicates that the Brown rat is forming part of the diet, as it usually does in the fox. A titre of 1:3000 to this serotype also indicates that perhaps that animal had a current infection. Nephritis is common in foxes (J. S. Fairley, personal communication) and leptospirosis may be one reason for its occurrence.

It is not unreasonable to imagine that infection in foxes results from their consuming infected prey. It would be unrealistic, though, to assume that leptospirosis is an important factor in the mortality of foxes since there would be little stability in a system in which the predator was highly susceptible to the parasites carried by the prey.

The absence of leptospirosis from a sample of Skomer voles for which full data were collected directed attention to island mammals and, although complete data are lacking for other species, good quality serum and histological material have been obtained. In the absence of dark-field data histology is a good second best, for the chances of finding histologically positive material are very good when the animal is excreting and the colonies are present in the kidney tubules. On Skomer the finding of antibodies to *L. canlcola* in two Great Black-backed gulls is of little significance as these birds feed to a great extent on the estuaries and streams of the mainland two miles away and infection may have been acquired in that way. It is nevertheless of some interest to find antibodies in a bird.

On the neighbouring island of Skokholm the finding of antibodies to *L. icterohaemorrhagiae* and *L. bataviae* poses an interesting question. The former serotype is most commonly, almost exclusively, carried by the Brown rat yet so far as is known the rat has never been found on the island. The island was, however, farmed for several years and so there is the possibility of original contamination by infected farm stock or dogs. The various possibilities of infection on this island have been reviewed by Twigg and Cuerden (1967). It has not been possible to demonstrate the presence of leptospirosis in the mice from Skokholm, voles from Skomer and various mammal species from Fair Isle and Rhum. There appears to be little leptospirosis on the islands but whether this is due to features of the habitat which have eradicated the organisms or to the fact that the first colonizers did not carry leptospirosis cannot be known. Adequate samples of *Rattus norvegicus* from several islands would be valuable in this problem.

There is, then, a large reservoir of leptospirosis in the British wild mammal fauna. Some species are clearly better maintenance hosts than others which have fleeting infections but, whether or not a species is a maintenance host, there is probably some time in its infection history, no matter how short, when it excretes leptospires and as such is contributing to the general contamination of soil and vegetation. When feeding and drinking, domestic animals must be at risk and from them, when infected, the infection of humans in close contact is the next link in the chain of infection. The Berkshire Downs survey has shown one facet of this picture.

The widespread occurrence of infection in wild mammals in Britain and the implications of this reservoir of leptospirosis to the well being of domesticated animals and man is a subject which deserves considerably more attention than has so far been accorded to it.

ACKNOWLEDGEMENTS

We should especially like to thank the Medical Research Council who have financed two of us (C.M.C. and D.M.H.) and provided funds for equipment and travelling. We are very grateful to Dr. L. H. Turner and Mr. R. J. Reed for advice on techniques and for their support; to Dr. A. McDiarmid, Mrs. M. Walters and Mr. P. Matthews of the Agricultural Research Council, Compton, for serum and assistance; to the Derwent Valley Water Board and Professor I. Chester Jones for facilities whilst collecting in Derbyshire; to Dr. R. J. Berry, Mr. H. G. Lloyd and Mr. R. Page for many island samples; to Mr. F. H. Grendon for the use of his woods and finally to the many people who have sent samples to assist the survey.

REFERENCES

Alston, J. M. and Broom, J. C. (1958). *Leptospirosis in man and animals*. E. and S. Livingstone, Edinburgh and London.

Ananin, V. V. (1954). Natural reservoirs of leptospirosis (title translated). *Zool. Zh. Ukr.* **33**, 331–340.

Borg-Petersen, C. (1944). *L. saxkoebing*, ein neuer serologischer Leptospiratyp. *Acta path. microbiol. scand.* **21**, 165–180

Broom, J. C. (1958). Leptospiral infection rates of wild rats in Britain. *J. Hyg., Camb.* **56**, 371–376.

Broom, J. C. (1959). *Infectious diseases of animals*, Diseases due to bacteria, Vol. 1. Stableforth, A. W. and Galloway, I. A. (eds.), Butterworths Scientific Publications, London.

Broom, J. C. and Coghlan, J. D. (1958). *Leptospira ballum* in small rodents in Scotland. *Lancet* 1041–1042.

Broom, J. C. and Coghlan, J. D. (1960). *Leptospira bratislava* isolated from a hedgehog in Scotland. *Lancet* 1326–1327.

Broom, J. C. and Gibson, E. A. (1953). Infection rates of *Rattus norvegicus* with *Leptospira icterohaemorrhagiae* in Great Britain. I. A rural area in Carmarthenshire, Wales. *J. Hyg., Camb.* **51**, 416–425.

Doeleman, F. P. J. (1932). Conjugal transmission of Weil's disease (title translated). *Ned. Tijdschr. Geneesk.* **76**, 5057–5058.

Elton, C., Ford, E. B. and Baker, J. R. (1931). The health and parasites of a wild mouse population. *Proc. zool. Soc. Lond.* **1931**, 657–721.

Ferris, D. H. and Andrews, R. D. (1967). Parameters of a natural focus of *Leptospira pomona* in Skunks and Opossums. *Bull. Wildl. Dis. Ass.* **3**, 2–10.

Kirk, R. (1938). Attempts to demonstrate leptospirosis in the Northern Sudan. *Trans. R. Soc. trop. Med. Hyg.*, **31**, 667–670.

McGuire, C. D. and Myers, D. M. (1957). Leptospirosis in Egypt. *Am. J. trop. Med. Hyg.*, **6**, 532–539.

Michna, S. W. (1958). *Leptospira pomona* antibodies in the sera of pigs: A preliminary report. *Vet. Rec.* **70**, 80–81.

Michna, S. W. (1967). Animal leptospirosis in the British Isles. *Vet. Rec.* **80**, 394–401.

Middleton, A. D. (1929). *Leptospira icterohaemorrhagiae* in Oxford rats. *J. Hyg., Camb.* **29**, 219–226.

Ostertag, H. (1950). *Leptospirosis icterohaemorrhagica* in Bulgarien. *Z. Hyg. InfektKrankh.* **131**, 482–500.

Seiler, H. E., Norval, J. and Coghlan, J. D. (1956). Leptospirosis in piggery workers. *Nature, Lond.* **177**, 1042.

Slavin, G., Salt, G. F. H. and Wolff, J. W. (1964). *Leptospira erinacei auriti* in a Water vole, *Arvicola amphibius*. *Nature. Lond.* **203**, 893.

Smith, C. E. G. and Turner, L. H. (1961). The effect of pH on the survival of leptospires in water. *Bull. Wld Hlth Org.* **24**, 35–43.

Trainer, D. O. (1961). Diseases of the Wisconsin White-tailed deer, *Odocoileus virginianus*. Ph.D. Thesis. University of Wisconsin, Madison.

Trainer, D. O., Hanson, R. P., Pope, E. P. and Carbrey, E. A. (1963). The rôle of deer in the epizootiology of leptospirosis in Wisconsin. *Am. J. vet. Res.* **24**, 159–167.

Twigg, G. I. and Cuerden, C. M. (1967). Leptospirosis on Skomer and Skokholm Islands. *Nature Wales* **10**, 154–158.

van der Hoeden, J. (1958). Epizootiology of Leptospirosis. *Adv. vet. Sci.* Academic Press, New York.

DISCUSSION

TAYLOR: Dr. Twigg has indicated a certain association between some of these serotypes and certain wild animals. It is interesting to speculate whether these organisms are harming the mammals in any way. I have worked on leptospirosis in rats and mongooses in the West Indies and, although this was a rather superficial survey, I formed the opinion that in some cases the animals, particularly the rats, were showing clinical evidence of the disease. I used to think I could spot a case in a rat by its hunched back and emaciated look; this was often associated with pitting of the kidney surface and, in some cases, adhesions between the kidneys and the mesentery. I would like to ask Dr. Twigg if he has noticed this sort of thing in any wild mammals he has looked at?

TWIGG: No, we have not, I am quite certain I cannot detect a field mouse infected with leptospirosis from a normal one nor can I detect any abnormality of the kidneys macroscopically. We are just commencing a study of the histology of these kidneys. I have been told by Dr. Fairley that there is a high degree of nephritis in foxes affected by leptospirosis in Northern Ireland.

MARSDEN: What proportion of brown rats would one expect to have an active leptospiral infection and how easy is it to reactivate latent leptospirosis with corticosteroids in this species?

TWIGG. I do not know the answer to the second question. So far as the first is concerned we have not looked at rats because the picture in those animals is reasonably well known. So far as I can recall other workers have found that the infection rate varies considerably from about 10% at the lowest to about 80% at the highest depending upon which time of year the samples are collected.

BLACKMORE: With regard to nephritis in foxes, a small survey we carried out several years ago showed 61% of our adult foxes had lesions and it was this same group which showed serological evidence of leptospirosis. About five different serotypes were involved. It appeared, therefore, that the interstitial nephritis was caused by the leptospires.

HAIG: Does water play a part in transmitting these infections? Do you think they are free-living in water and how long will they remain viable in this medium?

TWIGG: You can find leptospires in pond water but, of course, they are difficult to identify morphologically. Their pathogenicity would have to be checked in experimental animals such as hamsters or guinea pigs. When leptospires are excreted from a carrier animal they can last for considerable periods in contaminated water (at least 80 days has been recorded). It is interesting that although these micro-organisms live in the urine in the tubules, when shed in the urine they only live for a few hours unless the urine is diluted. The infection rates of rats in the Sudan, Egypt and Israel are very low indeed, sometimes practically non-existant; this is indicative of the fact that a wet environment does have some part to play.

TAYLOR: Do your percentages represent infection rates with the living organisms or do they represent contact rates based mainly on serological evidence?

TWIGG: The infection rate is based on the whole of the evidence. So far as the three common species are concerned, short-tailed vole, bank vole and long-tailed field mouse, there is a considerable amount of excretion. In the field mouse the excretion rate increases with age, whereas in the short-tailed vole most of the evidence of infection is serological and the number of excretors is low. In the bank vole it is the reverse, namely a fairly high rate of excretion with no evidence of antibody.

Symp. zool. Soc. Lond. (1968) No. 24, 99–105.

BRUCELLOSIS IN WILDLIFE

W. J. McCAUGHEY

Veterinary Research Laboratories, Stormont, Belfast, Northern Ireland

SYNOPSIS

The literature relevant to the incidence of Brucella infections in wild animals is reviewed. The aspects of infection in those animals which, because of their behaviour and habits, are likely to present a major threat to fully effective control and eradication in domestic animals are discussed in detail. The possibility of transfer of infection directly to man by this group of animals exists and some methods of prevention suggest themselves. The present situation in Northern Ireland in view of the eradication programme in progress has led to a limited amount of investigational work in this area. The results of these studies to date have proved uniformly negative. It is possible that this is influenced by the fact that, so far, only *Brucella abortus* biotypes have been isolated; also there is no serological evidence to indicate the presence of *Brucella melitensis* or *Brucella suis* biotypes.

In the eradication programme to date a large proportion of the reinfections in clear herds can be explained without reference to wildlife infection. The investigation into this disease in our wild animals is proceeding and may yet yield further evidence.

INTRODUCTION

In the present climate of concern over the incidence of this disease in domestic animals it is important to consider the possibility of reservoirs of infection remaining in our wild animals. The incidence of brucella infections in free-living animals have previously been reviewed by several workers including Stableforth and Galloway (1959) and McDiarmid (1962).

A wide variety of animal species are capable of becoming infected with brucella organisms. It may well be that all animals can be so infected if the challenge dose is high enough. There are definite species variations in susceptibility and this has been well recognized from the earliest investigation. Tunnicliff and Marsh (1935) found that elk grazing in an area also occupied by a heavily infected buffalo herd were much less heavily infected. There is, of course, a much more commonplace example of this phenomenon in domestic animals to be observed where our cattle and sheep flocks often graze over the same area. This was recognized as a possible hazard in the earliest days of the present eradication scheme in Northern Ireland but the examination of sera from large numbers of sheep taken from the whole country has been consistently negative for antibodies to brucella. In addition, cultural examinations of placentas and aborted foetuses from both sheep and pigs have proved negative for brucella organisms.

In this paper those wild animals which have been shown to be capable of infection are discussed with particular emphasis to any aspects which may appear of importance in these islands. To adopt these criteria leaves a wide range of suspect animals as it would appear that all, from the tiny dormouse to the various species of wild deer and wild pony, may have a part to play. It is possible, however, to narrow the field considerably by dealing with only those species which may be of greater importance because they co-habit with our grazing or housed stock. In this context the rat, the deer, the fox and the hare or rabbit appear to be potentially the most important.

INFECTION IN RATS

Infection in the common Brown rat (*Rattus norvegicus*) was first demonstrated by Hagan (1922). This was later substantiated by Bosworth (1937), who found that 5 out of 12 rats fed on brucella-contaminated materials yielded the organism at a later cultural examination; he also observed a very high correlation between serological titre and cultural recovery of the organism. The blood titres recorded, however, are much lower than those which have been observed in recently infected rats by W. J. McCaughey (unpublished data). Having demonstrated that the rat could become infected with *Brucella abortus* and that one focus of infection might be the kidney, Bosworth (1940) established an artificial colony of wild rats which he allowed to feed on material later fed to cattle. He was unable to demonstrate any transfer of infection to the cattle by this method and, in fact, he captured 167 wild rats from infected farms and proved that only one of these yielded the brucella organism. A low incidence in the wild rat has been confirmed by a number of workers including Menton (1937), Fitch and Bishop (1938) and McDiarmid (1951). There appears to be some discrepancy, however, as Karkadinovsky (1936) recovered *Brucella abortus* from 11 out of 34 rats captured from three infected premises. These workers were all dealing with *Brucella abortus* biotypes but a wide variety of other workers have demonstrated that *Brucella suis* and *Brucella melitensis* strains are probably more highly infective for these animals. It is, indeed, fortunate that neither of these strains appear to be endemic in the British Isles (Stableforth, 1960).

Experimental work carried out recently has shown that while the rat is relatively resistant to *Brucella abortus*, i.e., it requires a very heavy initial dose, it can become infected and can spread the organism through the urine. However, with the progress of eradication and the elimination of the reservoir of infection in our domestic animals it is unlikely that infection will persist in rats.

INFECTION IN DEER

Reports of infection in wild deer are of a much more widespread nature than those of infection in rats. The most relevant work is that carried out by McDiarmid (1951) when he demonstrated that in a sample of 80 deer from 3 different species 14% carried antibodies to brucella organisms. The postive sera in McDiarmid's survey belonged to japanese deer (*Sika nippon*) and fallow deer (*Dama dama*). Although McDiarmid's results were based purely on serological methods, other workers have produced cultural evidence of infection, recovering the organism from testicles, arthritic joints, eyes and foetuses. In Northern Ireland the position with regard to deer is highly favourable. There are only three well-established herds of deer and these are in areas which are not likely to be grazed by domestic animals. The examination of five sera during 1967 have proved negative.

It is reasonable to assume that, although to date there are no clinical reports of abortion caused by *Brucella abortus* in deer, this may occur. The reports of aborted foetuses mentioned by McDiarmid (1951) and Irvin (1966) in the British Isles would indicate that this may occur, although there are many other likely causes. A discussion with foresters working in areas where deer occur in Northern Ireland has not led to any reports of abortions being made. In Ireland there is one interesting husbandry anomaly shown in the Phoenix Park, Dublin, where 1760 acres of park-land are grazed concurrently by domestic animals, i.e., cattle and deer. Fortunately, the cattle are taken from the park in the autumn of each year for slaughter but once again there does not appear to be any real evidence of infection. No studies on these cattle have as yet been made.

INFECTION IN HARES

Reports of infection in hares deal mainly with *Brucella suis* biotypes and are largely confined to European countries. The first report is that of Witte (1941) in Germany and the latest that of Carlotto (1964) in Italy. Reports have also been received from Switzerland, France, Czechoslovakia, Poland, Denmark, U.S.S.R. and Hungary. There have been no reports from the British Isles. Thomsen (1959) produced a worldwide picture of the importance of brucellosis in hares based on a questionnaire which he sent to a large number of countries. He fully demonstrated that only in central and eastern European countries has there been any marked relationship between the infection in hares and the infection in pigs. In Africa, Asia, America and Australia no such relationship has been shown. In view of the importance of this condition and its widespread incidence (Bendtsen, Christiansen and Thomsen, 1956, 4% and Fritzsche, 1956, 6·8%), it is important to prevent the entrance of this biotype by the quarantine of imports and the blood testing of any herds from which animals have been presented for importation. This is particularly true of Ireland where the hare is a very ubiquitous animal. However, should the condition be imported it is unlikely that, under the present intensive management practised in pig herds in this country, contact between the two species will be very great.

In view of the recognized importance of this species and the fact that hares are extremely prevalent in Northern Ireland, 600 serum samples have been collected and tested for brucella antibodies. These have proved negative. In addition, this year, hare carcases from 4 different areas in the Province have been collected and examined culturally. No organisms resembling brucella have been recovered.

INFECTION IN WILD CARNIVORES

Agglutination tests from foxes in Argentina have shown that 11% were positive and that from these animals 8 cultures of *Brucella abortus* biotype 1 were recovered. The authors Szyfres and Tomé (1966) are unable to state definitely whether the brucella was naturally shed by the foxes or to what extent infected foxes could contribute to the dissemination of brucellosis. Szyfres and González Tomé (1967), however, were able to demonstrate that on 7 farms where animals and foxes were tested serologically, the percentage of foxes reacting at titres of 1/100 was more than that in the cattle. Over the last 5 years a small number of samples have been tested in Northern Ireland, with negative results, but in a recent survey 2 out of the first 5 foxes tested serologically have shown titres to both the serum agglutination test (S.A.T.) and the complement fixation test (C.F.T.). The exact significance of this is unknown but it is probable that these animals have been in contact with infection through acting as scavengers. This survey is being continued to see if the infection in this species can be tied up with a local infection in cattle. From experiences with domestic carnivores, in particular the greyhound, it would appear that persistent infection is unlikely to occur in this species.

Whilst most of the animals capable of transmitting the brucella organism do so by contaminating food with urine or discharges, there is also the possibility of transmission by common ectoparasites, e.g., the tick, biting fly and the ordinary house fly. It is certain that these insects can carry the organism. Wellmann (1951) demonstrated that *Brucella abortus* could be transmitted to guinea-pigs by the bites of *Stomxys calcitrans* and by a variety of other tabanids. In 1952 this worker further demonstrated that the infection would also be transmitted by *Musca domestica* (Wellmann, 1952) but he also demonstrated that stinging insects yielded a much higher percentage infection. A variety of ticks have been shown capable of transmitting brucella from one animal to another. There is inconclusive evidence, however, that brucella organisms can survive for any great length of time in these species. To date *Ixodes ricinus* has not been incriminated.

The importance of transmission by infected insects is likely to decrease as the sources of infection decrease.

In a brief paper of this kind it is impossible to deal with the complete picture with regard to the incidence of infection in wildlife. From observations carried out in Northern Ireland it appears that none of the species that are commonly accepted to be dangerous harbours the infection at present in this country. The finding of antibodies in the fox will take some time to further elucidate but it is indeed interesting. With the advancement of the eradication programme it is likely that the most common reservoir for infection, namely cattle, will disappear and that, with fewer herds infected, each one will have to be considered in more detail. The determination of the source of each fresh outbreak of contagious abortion will definitely provide much more information on this aspect. In conclusion it may be said that the British Isles are fortunate to have only *Brucella abortus* biotypes prevalent, that small rodents including the rat seem to be highly resistant to infection, that the hare and the rabbit are rarely affected with *Brucella abortus* and that although antibodies can be found in our deer, these are unlikely to be of great importance in the epidemiology of the disease. The major

importance of our wild animals would appear to be their activities as
scavengers, physically carrying contaminated material from infected
premises to a previously uninfected area.

REFERENCES

Bendtsen, H., Christiansen, M. and Thomsen, A. (1956). *Brucella suis* infection
in Hares as the cause of Enzootic Brucellosis in Pigs. *Nord. VetMed.*
8, 1–34.
Bosworth, T. J. (1937). The susceptibility of the wild rat to infection with
Brucella abortus—A preliminary note. *J. comp. Path.* **50**, 345–349.
Bosworth, T. J. (1940). Further observations on the wild rat as a carrier of
Brucella abortus. *J. comp. Path.* **53**, 42–49.
Carlotto, F. (1964). First outbreak of brucellosis in hares in Italy. *Vet. ital.* **15**,
962–963.
Fitch, C. P. and Bishop, L. M. (1938). The wild rat as a host of *Brucella abortus*.
Cornell Vet. **28**, 304–306.
Fritzsche, K. (1956). Brucellosis in wild hares in Rhineland-Palatinate. *Berl.
Münch. tierärztl. Wschr.* **69**, 301–307.
Hagan, W. A. (1922). The susceptibility of Mice and Rats to infection with
Bacillus abortus. *J. exp. Med.* **36**, 727–731.
Irvin, A. D. (1966). Some diseases of free-living wild mammals and their possible
relationship to human and domestic animal health. *Vet. Rec.* **79**, 776–
785.
Karkadinovsky, J. A. (1936). Sur la Brucellose des Rats gris savages. *C.r. Séanc.
Soc. Biol.* **121**, 1611–1613.
McDiarmid, A. (1951). The occurrence of agglutinins for *Br. abortus* in the blood
of wild deer in the South of England. *Vet. Rec.* **63**, 469–470.
McDiarmid, A. (1962). Diseases of free-living wild animals. *Agric. Stud.* No. 57.
F.A.O. p. 25.
Menton, J. (1937). Brucellosis in Man and Animals in Staffordshire: Laboratory
Observations. *Br. med. J.* Dec. 25. 1273–1275.
Stableforth, A. W. and Galloway, I. A. (1959). Infectious Diseases of Animals.
Butterworths Scientific Publications, London. Diseases due to Bacteria,
Vol. 1. p. 53–159.
Stableforth, A. W. (1960). Symposium on Brucellosis: 1. Introduction. *Vet.
Rec.* **72**, 419.
Szyfres, B. and Tomé, J. G. (1966). Natural Brucella infection in Argentine wild
foxes. *Bull. Wld Hlth Org.* **34**, 919–923.
Szyfres, B. and González Tomé, J. (1967). Natural brucella infection in wild
foxes in Argentina. *Boln Of. sanit. pan-am.* **62**, 144–150.
Thomsen, Axel. (1959). Occurrence of Brucella Infection in Swine and Hares,
with special regard to the European Countries. *Nord. VetMed.* **11**, 709–
718.
Tunnicliff. E. A. and Marsh, H. (1935). Bangs disease in bison and elk in the
Yellow-Stone National Park and on National Bison Range. *J. Am. vet. med.
Ass.* **86**, 745.
Wellmann G. (1951). Blood-sucking insects as mechanical carriers of Brucella
organisms. *Zentbl. Bakt. ParasitKde.* (Orig). **156**, 414–426.

Wellmann, G. (1952). Experimental transmission of brucellosis by insects. *Zentbl. Bakt. ParasitKde.* (Orig). **159**, 71–86.

Witte, J. (1941). Ueber das Vorkommen von Bangschen Abortus-Bakterien beim Wild in Freler Wildbahn. *Berl. Münch. tierärztl. Wschr.* No. 11. pp. 128–131.

Symp. zool. Soc. Lond. (1968) No. 24, 107–117.

PSEUDOTUBERCULOSIS IN FREE-LIVING WILD ANIMALS

N. S. MAIR

*Public Health Laboratory, Groby Road Hospital,
Leicester, England*

SYNOPSIS

Two hundred and twenty-two strains of *Pasteurella pseudotuberculosis* isolated from animals in Great Britian during the period 1961–1967 were examined at the Leicester public health laboratory. Of these, 44 strains (20%) were isolated from free-living wild animals, 18 from 5 species of mammals and 26 from 16 species of birds. The study confirmed previous observations that pseudotuberculosis in animals, as well as in man, is subject to a seasonal rhythm occurring more frequently in the winter when animals are exposed to cold and starvation. The principal reservoirs of infection are rodents and birds. The disease is transmitted to other animals, directly in the case of predators, and indirectly through the ingestion of contaminated feeding-stuffs. There is a close association between human and animal pseudotuberculosis. All the evidence points to contact with domestic pets as the major source of human infection. In recent years *Yersinia enterocolitica*, an organism closely resembling *P. pseudotuberculosis*, has been recognized as a cause of pseudotuberculosis in man and animals.

INTRODUCTION

Since 1961 strains of *Pasteurella pseudotuberculosis* have been submitted to the Leicester public health laboratory for serological typing. During the period 1961–1967, 222 strains of animal origin and 16 human strains were received from public health and hospital laboratories, veterinary investigation centres and zoological gardens. Of the 222 strains of animal origin, representing 57 different species, 44 were isolated from free-living animals, 67 from wild animals living in captivity, 57 from experimental animals and 54 from farm animals and domestic pets. The majority of strains (82%) belonged to serological type I, the type most prevalent in other European countries and found most frequently in man. A list of the affected free-living animals is presented in Table I. Of the 44 strains from free-living animals, 18 were isolated from 5 species of mammals and 26 from 16 species of birds.

TABLE I

Free-living animals infected with P. pseudotuberculosis

Mammals	No. of cases
Coypu (*Myocastor coypus*)	8
Fox (*Vulpes vulpes*)	1
Brown Hare (*Lepus europaeus*)	3
Rabbit (*Oryctolagus cuniculus*)	5
Field Vole (*Microtus agrestis*)	1
Birds	
Blackbird (*Turdus merula*)	1
Eider (*Somateria mollisima*)	1
Fieldfare (*Turdus pilaris*)	1
Hedge Sparrow (*Prunella modularis*)	1
House Martin (*Delichon urbica*)	3
Magpie (*Pica pica*)	1
Oystercatcher (*Haematopus ostralagus*)	1
Partridge (*Perdix perdix*)	3
Pheasant (*Phasianus colchicus*)	2
Pied Wagtail (*Motacilla alba*)	1
Redwing (*Turdus musicus*)	1
Swallow (*Hirundo rustica*)	1
Stock Dove (*Columba oenas*)	1
Tree Sparrow (*Passer montanus*)	1
Wood Pigeon (*Columba palumbus*)	6
Wren (*Troglodytes troglodytes*)	1
	44

Pseudotuberculosis in mammals

Hares

Pseudotuberculosis occurs most frequently in hares. Lucas (1961) examined 1678 hares over a 13-year period in France and found 16·5% infected with *P. pseudotuberculosis*. In Germany, Weidenmüller (1966) recorded an incidence of 12·6% over a 15-year period. In this country, Soltys found a high percentage of hares infected with *P. pseudotuberculosis* (Jamieson and Soltys, 1947). Clapham (1953) and McDiarmid (1962) also confirmed the presence of pseudotuberculosis in the brown hare *Lepus europaeus* and Macdonald (1962) quotes R. O. Muir as having diagnosed the disease in the scottish blue or mountain hare (*Lepus timidus scoticus*).

Foxes

Pseudotuberculosis is apparently rare in wild foxes. It would appear that only one case has been recorded in this country (Blackmore, 1964). The fox (*Vulpes vulpes*) in our series was one of many found dead near the Scottish border in Northumberland. A master of foxhounds had sent the carcase to the veterinary investigation centre in Newcastle-upon-Tyne in order to prove his theory that the dead fox was the victim of toxic chemicals used for the purpose of creating a "Silent Spring". However, characteristic necrotic nodules of pseudotuberculosis were present in all the viscera, including the lungs, from which *P. pseudotuberculosis* was isolated.

Coypu

The disease has been recorded in farm-bred coypus (Zschokke and Saxer, 1932; Claussen, 1934; Pilet, Valette, Labie and Fontaine, 1958; Oudar and Joubert, 1961) but there are few reports of pseudotuberculosis in the free-living animal. The 8 cases which we report here were diagnosed at the Coypu Centre in Norwich within the space of 2 years. Workers at this centre examined approximately 3000 coypus, live-trapped or otherwise caught and killed in Norfolk and Suffolk, over a 6-year period and found about 1% infected with *P. pseudotuberculosis* (R. M. Newson, personal communication). Details of the 8 cases are shown in Table II. According to Dr. Newson the poor condition of coypus in January 1963 was widespread and could not be correlated with disease: the animals were suffering from starvation and exposure. However, it is well known that pseudotuberculosis is latent in many animals and may become manifest under conditions of stress such as hunger, exposure or experimental inoculation.

Other mammals

On the Continent the disease has also been noted in roe-deer (*Capreolus capreolus*), mink (*Mustela lutreola*), marten (*Martes martes*), mole (*Talpa europaea*), hedgehog (*Erinaceus europaeus*) and marmot (*Marmota marmota*).

Pseudotuberculosis in birds

According to McDiarmid (1962), apart from hares, pseudotuberculosis is not a serious disease of free-living mammals. If this is true of

TABLE II

Details of coypus infected with P. pseudotuberculosis

Date of death	Age/sex	Weight (kg)	Serological type of organism	Notes
13/2/62	Ad. F.	3·70	IIA	Two circumscribed yellow lesions 5mm diam. on kidney containing green pus.
13/2/62	Ad. M.	4·20	IA	Multiple fine yellow lesions throughout liver.
21/1/63	6m. M	1·90	IIA	Multiple yellow lesions 3·4 mm diam. both kidneys: poor condition.
24/1/63	3m. M	0·95	IB	2mm abscesses on both kidneys: poor condition.
21/3/63	9m. M	3·30	IA	Large yellow nodule on one kidney: moderate condition.
21/6/63	8m. F	2·10	IA	Caught 11/6/63: kept at lab. and bitten by another coypu: died 21/6/63: *P. pstb.* isolated from purulent material at site of injury on foreleg.
12/2/64	2m. M	0·75	IA	Yellow lumps on kidney: spleen mottled, numerous small yellow spots on liver: poor condition: died in trap.
3/3/64	10m. F	4·05	IA	Multiple caseous 3mm lesions both kidneys: moderate condition.

mammals it is certainly not so of birds in which the disease is widespread. Clapham (1953) described an outbreak among stock doves (*Columba oenas*) in a small area of Hampshire during the winter of 1952. In Maryland, U.S.A., Clark and Locke (1962) recorded a major outbreak of the disease among Common Grackles (*Quiscalus quiscula*) during

March 1962. In a host list of wild birds affected by pseudotuberculosis, Macdonald (1962, 1965) mentions the following: Blackbird (*Turdus merula*), Coot (*Fulica atra*), Chough (*Pyrrhocorax pyrrhocorax*), Goldfinch (*Carduelis carduelis*), House Martin, (*Delichon urbica*) Jackdaw (*Corvus monedula*), Magpie (*Pica pica*), Oystercatcher (*Haematopus ostralagus*), Partridge (*Perdix perdix*), Pheasant (*Phasianus colchicus*), Pied Flycatcher (*Muscicapa hypoleuca*), Puffin (*Fratercula arctica*), Rook (*Corvus frugilegus,*) Stock Dove (*Columba oenas*), Starling (*Sturnus vulgaris*), Skylark (*Alauda arvensis*), Sparrowhawk (*Accipiter nisus*), Song Thrush (*Turdus ericetorum*), Swallow (*Hirundo rustica*), Swift (*Apus apus*), Tree Sparrow (*Passer montanus*), Wood Pigeon (*Columba palumbus*), Wren (*Troglodytes troglodytes*) and Willow Warbler (*Phylloscopus trochilus*). The additional species now shown to be affected include the Eider (*Somateria mollisima*), Fieldfare (*Turdus pilaris*), Hedge Sparrow (*Prunella modularis*), Pied Wagtail (*Motacilla alba*), and Redwing (*Turdus musicus*).

GEOGRAPHICAL DISTRIBUTION

Epidemiological studies indicate that pseudotuberculosis is essentially of European origin. It is widespread in France, Germany, the Low Countries, Switzerland, Great Britian, Scandinavia and European Russia. It is rare in the Mediterranean littoral, Africa, Asia and the New World.

SEASONAL VARIATION IN INCIDENCE

The disease is commoner in the winter months when animals, particularly free-living species, are exposed to cold and starvation. Table III shows the monthly distribution of 154 animal deaths and 101 human cases due to *P. pseudotuberculosis* in Great Britian during the period 1961–1967.

Deaths and cases are more numerous in the colder months of the year, November to April, reaching a maximum in January and showing a pronounced decrease in summer and early autumn.

ASSOCIATION BETWEEN HUMAN AND ANIMAL PSEUDOTUBERCULOSIS

It is evident that there is a close association between human and animal pseudotuberculosis. The clinical disease, however, runs a different course in man as compared to animals. Human pseudotuberculosis is localized, in general, to the mesenteric glands and usually

TABLE III

Monthly distribution of 154 animal deaths and 101 human cases

	Man	Animals
January	25	36 (14)
February	10	23 (7)
March	15	23 (5)
April	14	13 (3)
May	6	3 (3)
June	3	3 (1)
July	2	9 (3)
August	2	4 (0)
September	3	6 (0)
October	5	5 (0)
November	6	13 (5)
December	10	16 (3)
	101	154

Free-living animals in parenthesis

runs a benign course, whereas in animals the disease is characterized by pyaemia with foci scattered throughout the body and an invariably fatal outcome. *Post-mortem* small greyish white necrotic nodules are found in the spleen, liver and kidneys. Enlarged mesenteric glands are generally present. A fulminating septicaemia sometimes occurs in apparently healthy animals and results in death within 24 to 48 h. Pleural, pericardial or peritoneal exudates may be present and the organs appear congested without the occurrence of typical necrotic nodules. As we have already mentioned, besides definite clinical disease, *P. pseudotuberculosis* gives rise to a considerable amount of latent infection in animals.

RESERVOIRS OF PSEUDOTUBERCULOSIS

The principal reservoirs of infection are rodents and birds. Weidenmüller (1959) isolated *P. pseudotuberculosis* from 10 out of 30 field and house mice. Numerous enquiries in the U.S.S.R. (quoted by Bourdin, 1967) have established the presence of the organism in the Black Rat

(*Rattus rattus*), Brown Rat (*Rattus norvegicus*), House Mouse (*Mus musculus*), Striped Mouse (*Apodemus agrarius*), Wood Mouse (*Apodemus sylvaticus*), Common Vole (*Microtus arvalis*) and Water Vole (*Arvicola terrestris*). The organism is found in the faeces and urine of these animals without apparently causing any systemic disturbance. The mode of spread to other animals and man is illustrated in Fig. 1.

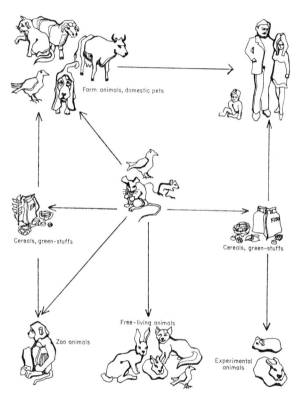

FIG. 1. Pseudotuberculosis cycle.

There are numerous reports of pseudotuberculosis affecting animals in zoological gardens. Fiennes (1966, 1968) has noted the appearance of a disturbing number of cases in the London Zoo in recent years and believes that the disease must now be regarded as endemic and likely to cause a number of losses each year. It is probable that much of the infection in zoological gardens is caused through contamination of food stores and food and water containers by rodents and birds. Paterson and Cook (1963) working at the Microbiological Research Establishment,

Porton, showed that *P. pseudotuberculosis* was repeatedly being introduced into their guinea-pig colony via the animals' greenfood. The main sources of infection were the huge flocks of wood pigeons which in Southern England are grossly affected with pseudotuberculosis. The pigeons void huge numbers of viable organisms in their excreta not only around their roosts but also as they feed on green crops such as kale, cabbage and brussel sprouts, which may be heavily contaminated. In cold weather the organisms survive in the faecal deposits on plants for several days.

It is probable that farm animals are also infected by contaminated feeding-stuffs. Blaxland (1947) described several outbreaks of pseudotuberculosis among flocks of turkeys in the winter of 1946. In two of the outbreaks, rats were trapped on the infected premises and in both cases an organism was isolated identical to that recovered from the turkeys. Predators among free-living animals are infected directly by rodents and birds and domestic cats are liable to infection for the same reason. Recently, Mair, Harbourne, Greenwood and White (1967) recorded the disease in two cats which were ferocious hunters and invariably ate the mice they caught.

It would appear that in the majority of cases man is infected by his domestic pets. Paul and Weltmann (1934) described the case of a labourer who died of pseudotuberculous septicaemia after working in a garden contaminated by the excreta of a cat suffering from diarrhoea. The animal was found to be infected with *P. pseudotuberculosis*. Randall and Mair (1962) investigating a family outbreak affecting four children found serological evidence of infection in their Corgi dog. In the only case of pseudotuberculous septicaemia recorded in this country Macaulay, Wilson, Abbott and Mair (1967) were able to demonstrate agglutinins to the infecting organism in the serum of a dog belonging to the patient. This dog which had a tendency to bite, had bitten the patient some days before he was taken ill. Cage-birds may also be a source of infection to man. Daniëls (1961) isolated *P. pseudotuberculosis* from the faeces of a 25-year-old man suffering from acute mesenteric adenitis and from the excreta of a canary owned by the patient. Infection may also occur, as it does in animals, through the ingestion of contaminated greenfoods. H. H. Mollaret (1967, personal communication) has noted from time to time a small explosion of cases among Parisiens in late spring and early summer. These patients are urban dwellers and have little, if any, contact with animals. They have, however, one thing in common—they enjoy salads made with dandelion leaves and the fields where they pick the dandelion leaves are known to be heavily manured with animal excreta.

INSECT VECTORS

The rôle of insect vectors remains uncertain. Blanc and Baltazard (1944) were able to show that the rat flea (*Xenopsylla cheopis*), when allowed to feed on an infected guinea-pig, remained a carrier for up to 35 days after the infected meal. Ul'yanova (1961) isolated 3 strains of *P. pseudotuberculosis* from 306 ixodid ticks collected from 80 cows in the outskirts of Leningrad. It is probable that insectiverous birds and mammals acquire the disease from insects which have fed on the droppings of infected rodents and birds.

YERSINIA ENTEROCOLITICA

It is not possible to conclude without some reference to an organism provisionally designated *Yersinia enterocolitica*. This organism possesses characteristics sufficiently distinct to separate it from *P. pseudotuberculosis*, yet resembles the latter sufficiently to justify its inclusion as a separate species in the genus *Yersinia*, the newly proposed generic name for *P. pseudotuberculosis* and *P. pestis*. Outbreaks of pseudotuberculosis due to *Y. enterocolitica* have been recorded in a pig farm in Algiers (Mollaret, 1966) and in chinchilla colonies in Switzerland (Becht, 1962), in Holland (Akkermans and Terpstra, 1963), in Germany (Knapp and Thal, 1963) and in Denmark (Frederiksen, 1964). Sporadic infection has been noted in the dog (Becht, 1962) and in hares (Mollaret, Chevalier and Deplanche, 1964; Mollaret and Lucas, 1965). In this country, the first isolation of *Y. enterocolitica* was made by McDiarmid in 1953 from the spleen of a hare found dead on the estate of the Agricultural Research Council at Compton in Berkshire. Recently we identified as *Y. enterocolitica* an organism isolated from a bush-baby (Galago), 1 of 3 which died after an outbreak of gastro-enteritis affecting 5 of 10 bush-babies in a private zoo. There is evidence that *Y. enterocolitica*, like *P. pseudotuberculosis*, is spreading in man and animals.

ACKNOWLEDGEMENT

I should like to thank Dr. R. M. Newson for permission to publish details of the coypus.

REFERENCES

Akkermans, J. P. W. M. and Terpstra, J. I. (1963). Pseudotuberculose bij chinchilla's veroorzaakt door een bijondere species. *Tijdschr. Diergeneesk.* **88**, 91–95.
Becht, H. (1962). Untersuchungen über die Pseudotuberkulose beim chinchilla. *Dtsch. tierärztl. Wschr.* **69**, 626–627.

Blackmore, D. K. (1964). A survey of disease in British wild foxes (*Vulpes vulpes*). *Vet. Rec.* **76**, 527–533.

Blanc, G. and Baltazard, M. (1944). Contribution à l'étude du comportement des microbes pathogènes chez les insects hématophages: la mémoire. *Archs Inst. Pasteur Maroc* **3**, 21–49.

Blaxland, J. D. (1947). Pasteurella pseudotuberculosis Infection in Turkeys. *Vet. Rec.* **59**, 317–318.

Bourdin, M. (1967) L'infection à bacille de Malassez et Vignal, zoonose d'avenir. Thèse veterinaire, Paris.

Clapham, P. A. (1953). Pseudotuberculosis among stock-doves in Hampshire. *Nature, Lond.* **172**, 353.

Clark, G. M. and Locke, L. N. (1962). Observations on pseudotuberculosis in common grackles. *Avian Dis.* **6**, 506–510.

Claussen, L. (1934). Über Bakteriämie durch das Bacterium pseudotuberculosis rodentium bei der Biberratte. *Dtsch. tierärztl. Wschr.* **42**, 243–247.

Daniëls, J. J. H. M. (1961). Enteral infection with Pasteurella pseudotuberculosis. Isolation of the organism from human faeces. *Br. med. J.* **2**, 997–998.

Fiennes, R. N. T.-W. (1966). The Zoological Society of London. Report of the Society's Pathologist for the year 1964. *J. Zool., Lond.* **148**, 363–380.

Fiennes, R. N. T.-W. (1968). The Zoological Society of London. Report of the Society's Pathologist for the year 1965. *J. Zool., Lond.* **154**, 173–191.

Frederiksen, W. (1964). A study of some Yersinia pseudotuberculosis-like bacteria ("Bacterium enterocoliticum" and "Pasteurella X"). *Proc. XIV Scand. Congr. Path. Microbiol.* Oslo, 1964. Oslo Universitetsforlaget. 1964. 103–104.

Jamieson, S. and Soltys, M. A. (1947). Infectious Epididymo-orchitis of Rams Associated with Pasteurella pseudotuberculosis. *Vet. Rec.* **59**, 351–353.

Knapp, W. and Thal, E. (1963). Untersuchungen über die kulturellbiochemischen, serologischen, tierexperimentellen und immunologischen Eigenschaften einer vorläufig "Pasteurella X" benannten Bakterienart. *Zentbl. Bakt. ParasitKde.* I. (Orig.), **190**, 472–484.

Lucas, A. (1961). Quelques maladies du gibier française transmissibles à l'homme. *Concours. méd.* **83**, 3627–3656.

Macaulay, J. D., Wilson, J. A. C., Abbott, J. D. and Mair, N. S. (1967). Fatal Case of Pasteurella pseudotuberculosis Infection Associated with Hepatic Cirrhosis. *Br. med. J.* **2**, 553–554.

McDiarmid, A. (1962). Diseases of free-living wild animals. *F.A.O. agric. stud.* No. 57.

Macdonald, J. W. (1962). Mortality in wild birds with some observations on bird weights. *Bird Study* **9**, 147–167.

Macdonald J. W. (1965). Mortality in wild birds. *Bird Study* **12**, 181–195.

Mair, N. S., Harbourne, J. F., Greenwood, M. T. and White, G. (1967). Pasteurella pseudotuberculosis infection in the Cat: two cases. *Vet. Rec.* **81**, 461–462.

Mollaret, H. H. (1966). L'infection humaine a "Yersinia enterocolitica". *Path. Biol. Paris.* **14**, 981–990.

Mollaret, H. H., Chevalier, A. and Deplanche, M. C. (1964). Contribution à l'étude d'un nouveau groupe de germes proches du bacille de Malassez et Vignal. 1. Caractères culturaux et biochimiques. *Annls Inst. Pasteur, Paris.* **107**, 121–127.

Mollaret H. H. and Lucas, A. (1965). Sur les particularités biochimiques des souches de Yersinia enterocolitica isolées chez les lièvres. *Annls Inst. Pasteur, Paris*, **108**, 121–125.

Oudar, J. and Joubert, L. (1961). Trois nouveaux cas de pseudotuberculose à "Cillopasteurella pseudotuberculosis" chez le ragondin (*Myocastor coypus Molina*). *Bull. Soc. Sci. vét. Lyon* **63**, 313–318.

Paterson, J. S. and Cook, R. (1963). A method for the recovery of Pasteurella pseudotuberculosis from faeces. *J. Path. Bact.* **85**, 241–242.

Paul, O and Weltmann, O. (1934). Pseudotuberkulose beim Menschen. *Wien. klin. Wschr.* **1934**, 603–604.

Pilet, Ch., Valette, L., Labie, Ch. and Fontaine, M. (1958). Pseudotuberculose dans un élevage de Myocastors. *Bull. Acad. vét. Fr.* **31**, 299–303.

Randall, K. J. and Mair, N. S. (1962). Family outbreak of Pasteurella pseudotuberculosis infection. *Lancet.* **1**, 1042–1043.

Ul'yanova, N. I. (1961). The natural focal occurrence of pseudotuberculosis. *J. Microbiol. Epidem. Immunobiol.* **32**, 2268–2273. (Translated from Russian).

Weidenmüller, H. (1959). Zur Rodentiose bei Tier und Mensch. *Tierärztl. Umsch.* **8**, 256–259.

Weidenmüller, H. (1966). Pseudotuberkulose bei Wildtieren. *Tierärztl. Umsch.* **21**, 447–448.

Zschokke, W. and Saxer, E. (1932). Beobachtungen bei Pelztierkrankungen. Ein Fall von Pseudotuberkulose beim Sumfbiber (Nutria). *Schweizer Arch. Tierheilk.* **74**, 446–450.

Symp. zool. Soc. Lond. (1968) No. 24, 119–131.

MYCOBACTERIAL INFECTIONS IN FREE-LIVING WILD ANIMALS

J. DEANS RANKIN and A. McDIARMID

Agricultural Research Council, Institute for Research on Animal Diseases, Compton, Newbury, Berkshire, England

SYNOPSIS

For many years cases of mycobacterial infection have been reported in a wide variety of wild animals. Some of these infections were typical mammalian or avian tuberculosis but a number have been atypical and some have not, so far, been typed.

The virtual eradication of tuberculosis from the cattle herds and poultry flocks in Britian has increased the relative importance of such infections in our wildlife population. Attention is drawn to a number of these infections and their possible aetiology and implications are discussed.

INTRODUCTION

There are two obvious ways in which we can deal with this subject. We can consider each species of acid-fast bacillus and list the various animal in which it has been found or, alternatively, we can consider the animal and examine the mycobacteria usually associated with it. We have chosen the second method because we feel that it is the host-parasite relationship which is important and, equally important, the relationship of one animal species with another.

Only the more important animal species have been considered and, because of the limited time available, the cold-blooded animals and their specific mycobacteria have been excluded. Those interested in the mycobacteria of the cold-blooded animals should consult the review article by Vogel (1958).

MONKEYS

It has long been recognized that tuberculosis is the major killing disease of monkeys in captivity. These animals are susceptible to both human and bovine types of the bacillus and the rate of spread within a population is very rapid. Six months after the introduction of the disease 50% of a colony will be infected and 3 months later it will be difficult to find a monkey not reacting to a tuberculin test. The spread within the individual body is also rapid—the time from infection to death being from 4 to 10 weeks. Thus it is not uncommon to lose 10% of a colony within a period of 3 months.

These observations are based on data collected from various zoological collections and from monkey colonies attached to drug companies and research laboratories. There is, however, little evidence of free-living monkeys being infected with tuberculosis. Cobbett, 1917; Vallée and Panisett, 1920; Nieberle, 1932 and Urbain, 1941 have each remarked on the absence of the disease after many hundreds of post-mortem examinations. However, Griffiths (1939) observed calcarious lesions in a number of rhesus monkeys sent from India to a laboratory in England and it is perhaps worth noting that this species can often be seen around the temples and close to human habitation where they scavenge for food. It seems probable that, although highly susceptible to tuberculosis, the monkey is safe until it comes directly or indirectly into contact with man.

In this respect it is interesting to note that when rhesus monkeys began to be imported in large numbers into this country for polio vaccine production, tuberculosis was a very serious problem. Within a year or two the situation improved very markedly. This can be explained if we consider that the first monkeys to be caught up were those around the villages and towns whereas later, when all the local monkeys had been captured, the supply was coming from the forests.

Fremming, Benson and Young (1955) tuberculin-tested a group of monkeys captured in the outskirts of New Delhi and reported a reactor rate of 50% whereas only 0·5% of a group captured deeper in the forest reacted.

VOLES

In 1937 Wells described a tuberculous condition affecting field voles in their natural habitat. He isolated from the lesions a mycobacterium quite different from the type species. This new acid-fast bacillus was very slow growing and growth was poor in the presence of glycerine. The most striking feature of this organism was its pleomorphism. Curved forms abounded and crooked, sickle, spiral and S-forms were all present. Wells names the bacillus *Mycobacterium tuberculosis* var. *muris*. Its more common name is the Vole Acid-fast Bacillus. This bacillus proved to be considerably less virulent than the other mammalian types and for a time it was thought that it might form a useful alternative to B.C.G.

as a vaccine strain. Further work, however, revealed that it was not entirely safe and occasionally produced certain unfortunate side effects when used to vaccinate rabbits and guinea-pigs.

Wells examined over 4000 voles and found lesions in 20% of them, the percentage varying from 0–66% in different areas. It appears that the normal range of a vole family seldom exceeds 15 yards and that areas of high and low morbidity could occur almost within the same field.

In due course as shown in Table I other species of small wild mammals were also shown to be affected but true tuberculosis was never found.

TABLE I

Incidence of Mycobacterium tuberculosis *var.* muris *in 4 species*

Species	No. examined	Percentage affected
Field Vole (*Microtus agrestis*)	4000	20
Bank Vole (*Clethrionomys glareolus*)	223	10·3
Shrew (*Sorex araneus*)	550	1·4
Wood Mouse (*Apodemus sylvaticus*)	175	2·3

(Wells, 1946)

The distribution of the lesions suggested a route of infection other than pulmonary. The lungs were affected in only 11% of cases whereas 54% of cases showed mesenteric involvement. In a considerable number of cases only the superficial lymph nodes of the head and neck were infected. Wells concluded that the principal method of infection was by ingestion of diseased voles or by the infection of wounds received during fighting. He confirmed this experimentally by observing a colony into which he had introduced a diseased vole.

RATS

The wild brown rat (*Rattus norvegicus*) is known to be the occasional victim of tuberculosis. Usually the lesions are associated with the alimentary tract and are indicative of oral infection. Both the avian and bovine type have been isolated depending on the habitat of the rat. In Denmark, Plum (1942) has recorded a 16·5% incidence of avian type infection in rats but says that in heavily infected localities up to 70% can be infected.

Like the vole, the rat also appears to have a species of mycobacterium of its own. This is the Rat Leprosy Bacillus (*Mycobacterium lepraemurium*). It was first described at the beginning of the century by Stefansky (1903) who saw it in material from rats suffering from a leprosy-like disease. The incidence is usually around 1% but a 5% incidence has been reported. The bacillus is strongly acid-alcohol-fast and although it appears in very large numbers in infected tissues it is difficult to maintain in culture. It is a separate species from that causing leprosy in man.

The disease manifests itself in two ways—a glandular form and a musculo-cutaneous form. In neither case is there any death of tissue comparable to the necrosis and caseation of tuberculosis but there is the usual infiltration of the tissues with large mononuclear cells. These are rich in cytoplasm and are usually packed with acid-fast bacilli.

The manner in which natural infection occurs is not known but successful experimental transmissions have occurred after intradermal, subcutaneous and intraperitoneal inoculation.

DEER, ANTELOPE AND OTHER RUMINANTS

Among the large mammals tuberculosis has been encountered most frequently in the various species of deer. This may be because of the interest taken in deer by sportsmen in various countries, especially in Germany where facilities are often available for further laboratory examination of diseased tissues.

Records from Germany include cases of acute miliary tuberculosis in roe deer (*Capreolus capreolus*) (Schmidt, 1938; Platen, 1939 and Hillenbrand, 1940). The disease has also been described in red deer (*Cervus elaphus*) by Witte (1940). Christiansen (1931) and Hermansson (1943) have also described the disease in the same species in Denmark. The latter infections were of the avian type and it is interesting to note that Hopkinson and McDiarmid (1964) also encountered generalized avian tuberculosis in red deer in Scotland. Subsequent investigations, which are continuing, have shown a high recovery rate of mycobacteria in culture from ileo-caecal lymph glands from different species of British deer—the highest so far being in roe deer (27·0%), the lowest in red deer (5·0%) and the incidence in fallow deer (*Dama dama*) varying according to the locality; for example 13·0% in the New Forest. Frank clinical cases in all these species have been rare and the various varieties of acid-fast organisms which we have isolated have not yet been finally identified.

Bouvier, Burgisser and Schweizer (1951) reported a case of bovine tuberculosis in a chamois (*Rupicapra tragus*). The lesions were miliary and affected both lungs. This was the first case of tuberculosis the authors had seen in 132 chamois shot. Bouvier had also encountered a single case in 199 roe deer killed. From these figures the authors concluded that tuberculosis was uncommon in the game animals of Switzerland.

Tuberculosis has also been encountered in Africa in antelopes (Robinson, 1953) and in buffalo (Guilbride, Rollinson, McAnulty, Alley and Wells, 1963). Of 13 wild buffalo shot in Uganda, 8 were tuberculous. All the infections were of bovine type. A case has also been reported in a kudu from Cape Province, S. Africa (de Kock, 1938).

The absence of tuberculosis in African game animals has been commented upon by R. Sachs (1968, personal communication). During a period of about 4 years in the Serengeti National Park area of Tanzania he carried out detailed post-mortem examinations on more than 500 antelope without detecting any evidence of tuberculosis.

An acid-fast bacillus commonly associated with, or perhaps even confined to, the domesticated ruminant is *Mycobacterium paratuberculosis* (the Johne's bacillus) and it might be presumed that infection with this organism would also occur in wild ruminants. Although there are numerous recorded instances of Johne's disease affecting deer and other wild ruminants kept in parks or in zoological gardens where there has been the possibility of contact with domesticated stock, such as cattle or sheep, there is no evidence that such infections occur in free-living wild animals, apart from the fact that many of the mycobacteria isolated by one of us (A. McD.) from deer require mycobactin for growth and are, so far, indistinguishable culturally from *M. paratuberculosis*. With the co-operation of Dr. Sachs of the Serengeti Veterinary Research Unit, one of us (J.D.R.) has examined specimens from 67 animals from 11 different species* grazing the Serengeti and no evidence of paratuberculosis was found.

Wildebeeste (*Connochaetes taurinus*) (28), Thomsons Gazelle (*Gazella thomsonii*) (11), Grants Gazelle (*Gazella granti*) (7), Eland (*Taurotragus oryx*) (4), Kongoni (*Alcelaphus buselaphus cokei*) (4), Dik-Dik (*Rhynchotragus* sp.) (4), Buffalo (*Synceros caffer*) (3), Waterbuck (*Kobus defassa*) (2), Masai Sheep (2), Impala (*Aepyceros melampus*) (1), Topi (*Damaliscus korrigum*) (1).

TABLE II

Some wild birds in which tuberculosis has been reported

Farmland birds		Birds of prey		Game birds	
Crow (*Corvus brachyrhynchos*)	10·0%	Golden Eagle (*Aquila chrysaetos*)	(3)	Pheasant (*Phasianus colchicus*)	(13)(2)(1)
Rook (*Corvus frugilegus*)	1·0%	Kestrel (*Falco tinnunculus*)	(1)	Capercaillie (*Tetrao urogallus*)	(1)
Jackdaw (*Corvus monedula*)	1·0%	Sparrow Hawk (*Accipiter nisus*)	(1)	Black Grouse (*Tetrao tetrix*)	(1)
Starling (*Sturnus vulgaris*)	5·0%	Brown Hawk (Australian) (*Falco*		French Partridge (*Alectoris rufa*)	(1)
House Sparrow (*Passer domesticus*)	5·0% to 40·0%	*berigora*)	(1)	Grey Partridge (*Perdix perdix*)	(1)
Woodpigeon (*Columba palumbus*)	1·0% to 6·0%	Barn Owl (*Tyto alba*)	(1)		
Golden Plover (*Pluvialis apricaria*)	(1)				
Green Plover (*Vanellus vanellus*)	(1)				
Blackbird (*Turdus merula*)	(1)				

Sea birds		Wildfowl	
Herring Gull (*Larus argentatus*)	(5)	Grey Teal (*Querquedula gibberifrons*)	(1)
Black-headed Gull (*Larus ridibundus*)	(1)	Mallard (*Anas platyrhynchos*)	(1)
Cormorant (*Phalacrocorax carbo*)	(1)	Wigeon (*Anas penelope*)	(1)
Eider Duck (*Somataria mollissima*)	(1)	Greylag Goose (*Anser anser*)	(1)
Oyster Catcher (*Haematopus ostralegus occidentalis*)	(1)	Pink-footed Goose (*Anser fabalis*)	(1)

(Numbers in brackets refer to isolations)

BIRDS

Those who keep captive wild birds will know that tuberculosis can be a common cause of death and that once introduced to the aviary the mortality rate is seldom insignificant and sometimes reaches 100%. It is more difficult to assess the extent of the disease in the free-living bird population but it would appear that the pattern may be similar to that already referred to in monkeys, viz. that the disease is only common where birds have a close association with man and his poultry. There are, however, certain exceptions which will be referred to later.

In a survey of post-mortem examinations involving some 513 birds from 74 different species, Keymer (1958) found tuberculosis in a single pheasant (*Phasianus colchicus*). The majority of reports of tuberculosis in free-living birds refer to single or sporadic isolations and it is only in the crow, starling, sparrow, wood pigeon and pheasant that there is any suggestion of enzootic disease. Table II has been compiled mainly from a review by Wilson (1960) but we have included some of our own information.

It can be seen that the greatest number of reports refer to crows, starlings, sparrows, wood pigeons and pheasants all of which have some contact with farms and therefore with domestic poultry. It is easy to visualize the spread of infection from these common birds to the birds of prey. The others (gulls and wild fowl) are probably chance infections resulting from scavenging or from a contaminated environment.

The infection in wood pigeons must be considered in a little greater detail. A proportion of these undoubtedly result from infection with the classic avian type bacillus but many birds are found with gross lesions of tuberculosis teeming with acid-fast bacilli which are difficult, or impossible, to cultivate. The best results have been obtained when the medium has contained an accessory growth factor (mycobactin) and the tubes have been incubated for 3 to 6 months (McDiarmid, 1962). The resulting growth has had the characteristics of some of the rough unclassified or so-called "anonymous" or "opportunist" strains. Serial passage in poultry, however, has resulted in a reversion to the smooth fully virulent avian type. The lesions in pigeons are always associated with the alimentary tract and we consider that, because of the method of feeding of the squabs (young pigeons), this is now likely to be a self-perpetuating disease in pigeons, whatever its original source.

DISCUSSION

Wilson (1960) stated that tuberculosis in wild birds is likely to be the result of direct or indirect contact with domestic poultry and that it is not likely to form a reservoir of infection for the domestic fowl. When one considers the pattern of tuberculosis in the monkey, Bouvier's observations on the chamois and roe deer in Switzerland and Sachs experience in the free-living wild ruminants of the Serengeti, one might even be justified in assuming that all tuberculosis in free-living wild-life has its origin in man and his domesticated animals.

This may very well have been true in the past but is it still true today? When many of the observations quoted in this paper were being made the incidence of tuberculosis in our national poultry flock was over 10%; by 1950 this had been reduced to 3% and today it is around 1%. So we now have the situation where the incidence in some of our wild birds is 5 to 10 times that in our domestic poultry whereas previously it was much the same.

This potential source of infection is not likely to be a hazard to poultry kept under modern intensive systems of management but it could very well be a serious menace to poultry living under semi-intensive or free-range conditions.

Also, in spite of the spectacular decrease in tuberculosis in poultry flocks, there has been no corresponding decrease in the large number of pig heads condemned in our abattoirs because of infection with tuberculosis mainly of the avian type. A single enterprise has quoted losses in the region of £30,000 per annum from this wastage. One is forced to the conclusion that the source of this infection must be in the large flocks of starlings and sparrows which congregate around the feeding troughs or the contamination of the grass by flocks of wood pigeons.

Lastly the incidence of avian sensitivity in our cattle population remains high and Smith (1955) has shown a correlation between this and the areas where birds congregate. While frank disease in cattle due to the avian bacillus is not common, it is by no means as rare as we once believed. In any case widespread sensitivity to avian tuberculosis can cause considerable difficulties in the interpretation of tuberculin tests in cattle, particularly in those countries where only mammalian tuberculin is used.

Our recent investigations in British deer indicate a high carriage rate of acid-fast organisms mainly of the typical or atypical avian types; this is, in itself, an extremely interesting situation as this might well be a unique opportunity for the avian strains to be passaged through a mammalian host owing to the fact that deer in certain areas, in order to supplement their mineral requirements, chew bones of members of their own or related species dying in remote parts of the countryside. Meat inspection of game is almost non-existent and the opportunity for infection must occasionally be presented to those handling this material. While it is unlikely that mycobacterial infections in free-living wild animals will be a major threat to our agricultural industry, it is necessary that we should be aware of these sources of potential infection and that we should take them into consideration when we are designing management systems for our domestic stock and investigating break-downs in their tuberculosis status.

REFERENCES

Bouvier, G., Burgisser, H. and Schweizer, R. (1951). Tuberculose chez un chamois. *Schweizer Arch. Tierheilk.* **93**, 689.

Christiansen, M. (1931). Ausgedehnte Organtuberkulose bei einem Reh, verursacht durch den Geflügeltuberkelbazillus. *Z. InfektKrankh. parasit. Krank. Hyg. Haustiere.* **40**, 165.

Cobbett, L. (1917). The causes of Tuberculosis. Cambridge.

de Kock, G. (1938). Wild animals as carriers of infection. *S. Afr. med. J.* **12**, 725.

Fremming, B. D., Benson, R. E. and Young, R. J. (1955). Procurement of monkeys for the radio-biological laboratory. *Scient. Mon., N.Y.* **80**, 260.

Griffiths, A. S. (1939). Infections of wild animals with tubercle bacilli and other acid-fast bacilli. *Proc. R. Soc. Med.* **32**, 1405–1412.

Guilbride, P. D. L., Rollinson, D. H. L., McAnulty, E. G., Alley, J. G. and Wells, E. A. (1963). Tuberculosis in the Free-Living African (Cape) Buffalo. *J. comp. Path.* **73**, 337.

Hermansson, K. A. (1943). A Case of Tuberculosis in an Elk. *Skand. VetTidskr.* **33**, 66.

Hillenbrand, F. (1940). Pathologisch-anatomische Befunde der Tropenkrank-heiten des Pferdes in unseren Afrikanischen Kolonien. Inaug. Diss. Leipzig, p. 28.

Hopkinson, F. and McDiarmid, A. (1964). Tuberculosis in Free-Living Red Deer, *Cervus elaphus*, in Scotland. *Vet. Rec.* **76**, 1521.

Keymer, I. F. (1958). A Survey and Review of the Causes of Mortality in British Birds and the Significance of Wild Birds as Disseminators of Disease. *Vet. Rec.* **70**, 713.

McDiarmid, A. (1962). Diseases of Free-Living Wild Animals. *F.A.O. agric. Stud.* No. 57.

Nieberle, K. (1932). Tuberkulose bei Tien. *Ergeben. allg. Path. path. Anat.* **25**, 631.

Platen, J. (1939). Ein Fall von akuter Miliartuberkulose bei einem Reh. *Berl. Münch. tierärztl. Wschr.* **55**, 536.

Plum, N. (1942). Studies on the Occurrence of Avian Tuberculosis among Wild Birds especially Gulls and Sparrows, and in Rats and Hares. *Skand. vetTidskr.* **32**, 465.

Robinson, E. M. (1953). A few Cases of Tuberculosis. *Jl S. Afr. vet. med. Ass.* **24**, 97.

Schmidt, H. W. (1938). Tuberculosis in the Roe Deer. *Dt. tierärztl. Wschr.* **46**, 482.

Smith, A. (1955). Observations on the incidence of non-specific sensitivity in cattle. Thesis. Roy. Coll. vet. Surgeons, London.

Stefansky, W. K. (1903). Eine lepraähnliche Erkrankung der Haut und der Lymphdrüsen bei Wanderratten. *Zentbl. Bkt. ParasitKde* **33**, 481.

Urbain, A. (1941). Tuberculosis in Captive Monkeys. *Bull. Acad. Méd.* **124**, 281.

Vallée, H. and Panisett, J. (1920). Les Tuberculoses Animales. Paris, p. 528.

Vogel, H. (1958). Mycobacteria from Cold-Blooded Animals. *Am. Rev. Tuberc. pulm. Dis.* **77**, 823.

Wells, A. Q. (1937). Tuberculosis in Wild Voles. *Lancet* **232**, (1), 1221.

Wells, A. Q. (1946). The murine type of tubercle bacillus (the vole acid-fast bacillus). *Spec. Rep. Ser. med. Res. Coun.* No. 259.

Wilson, J. E. (1960). Avian Tuberculosis. *Br. vet. J.* **116**, 380.

Witte, J. (1940). Tuberkulose beim Rotwild in freier Wildbahn. *Berl. Münch. tierärztl. Wschr.* **56**, 349.

DISCUSSION

IRVIN: There have been a few cases of infection in cats in the Netherlands with the vole bacillus. Have any similar cases been recorded in this country?

RANKIN: I have not heard of any.

BRAMLEY: With reference to avian tuberculosis in deer, I have found two cases in red deer in Westmorland. On post mortem examination, extensive grape-like lesions were found affecting the thoracic, abdominal and pelvic cavities. Both cases also showed abnormalities of the antlers. Can you tell me what you have found in red deer in Scotland. Were the lesions similar?

RANKIN: Yes, they were similar but perhaps Dr. McDiarmid would like to enlarge on this.

McDIARMID: The lesions we found in the Scottish red deer were just as you described. The first case was extremely interesting because the animal was actually seen alive. It was stalked in the usual manner and shot because it was thought that it had been wounded previously. It was a very weak looking animal with a very poor head, the antlers being about half the size they should have been. On opening the carcase, lesions were found throughout the thoracic and peritoneal cavities involving the parietal pleura and peritoneum.

DELL: Have rapid growing acid-fast organisms been isolated from wild animals in this country?

RANKIN: The only case I have encountered was in a crocodile from this Zoological Park. The colonies of the organism were highly pigmented, orange to red in colour and grew in two or three days. It was, I believe, a pulmonary infection but this is going back some twenty-odd years. We transmitted the infection back to young crocodiles which we obtained from here.

ROBERTS: We have been working on avian tuberculosis in pigs and in the course of this we have found reference in the literature to a reported absence of avian tuberculosis in Queensland and Finland. Such caseous lesions as do occur are considered there to be due to *C. equi* infection. In Scotland all of these infections are most definitely avian tuberculosis. Much of our material comes from Aberdeenshire and I doubt if there is much in the way of financial loss as all they do is to remove the affected submaxillary lymphatic gland. In view of the numerous reports on the danger of avian tuberculosis to man, could the infection of pig head glands be an important risk for humans?

RANKIN: I cannot really say that avian tuberculosis does not occur in the countries you mentioned. It is not always easy to culture the organism from a pig head gland although one may observe the bacilli in smears and in sections. A proportion of these strains will, in fact, grow on ordinary egg media but we have found that the recovery rate of these organisms can be increased by adding an accessory factor such as mycobactin to the medium and incubating for a much longer period. Some of the strains take months to grow.

CHAIRMAN'S SUMMING-UP

Most of what has been said this afternoon has justified the remarks I made at the beginning, that animals, in general, do not suffer from most infections unless they congregate in large groups and that man by forcing his farm animals into large groups has clearly increased infections among them and in so doing has started up infections in animals outside that would probably not have occurred had he not accumulated animals himself. The other interesting thing, I think, is that pseudotuberculosis appears to be basically a disease of birds and the spread of the disease among animals other than birds seems to be largely due to the habit of birds fouling food. I think what has been stated throughout is a further justification for my plea for behaviour examinations of animals. Man, by his own behaviour, has increased the opportunities for animals to become infected; very frequently it seems to me the one thing left out of serious consideration is the behaviour of man himself. You will have noticed I think, through most of the papers, a gently moralistic tone and I would of course be quite unwilling not to indulge in this myself. The fact that surgeons drop things from which bacteriologists are expected to derive cultures, for instance, is one of these things I am just as familiar with as Dr. Mair.

Brucellosis seems to me to be a very strange kind of disease; it seems so limited to cattle and men although other animals are possibly capable of being infected. I think it is worth remembering that animals may be infected in the laboratory with large concentrations of organisms but they may be difficult to infect under field conditions when presented with a small concentration. I think Dr. Taylor's remarks on animals becoming infected in rivers contaminated with sewage is a good example; animals that might not readily be infected at all under natural conditions are very liable to be infected if presented with vast numbers of organisms in a comparatively small volume of water. Those of you familiar with the vast amount of sewage discharged into the English Channel will hardly be surprised at this. Even the volume of water available there is inadequate to dilute, over the coastline at any rate, the number of organisms normally discharged into that area.

PROTOZOA

CHAIRMAN: PROFESSOR P. C. C. GARNHAM

CHAIRMAN'S INTRODUCTION

The following four papers include the three most important protozoa to be found in the blood of animals, trypanosomes, piroplasms and malaria parasites. These protozoal infections are all zoonoses to a greater or lesser extent; the third human case of babesisis has just been described from Ireland. They are rare as human zoonoses and even Rhodesian sleeping sickness is probably not acquired as much from animals as from human infections.

We heard yesterday about the effect of stress (i.e. exposure to abnormal conditions of all sorts) on wild animals in captivity and how this factor enhances a latent infection; there is no doubt that some animals in captivity may rapidly die of a protozoal infection which scarcely affects them in nature. I think, for instance, of *Lankesterella* in fledgeling sparrows or *Toxoplasma* in gondi, mink and armadillos but, as far as I know, the trypanosomiasis of wild game still remains a mild condition when the animals are confined in a zoo.

Symp. zool. Soc. Lond. (1968) No. 24, 133–145.

TRYPANOSOME INFECTIONS OF WILD MAMMALS IN RELATION TO TRYPANOSOME DISEASES OF MAN AND HIS DOMESTIC STOCK

E. A. WELLS and W. H. R. LUMSDEN

Applied Protozoology Research Unit,
Royal (Dick) School of Veterinary Studies,
University of Edinburgh, Scotland

SYNOPSIS

Trypanosome infections are widely distributed among mammals both taxonomically and geographically. The infections seem to be typically benign and pathogenicity to be associated with hosts not usually involved in natural transmission cycles. When these hosts happen to be man or his domestic animals very important disease situations may be provoked. There are many colloquial terms for these diseases but they may be grouped under a few of the most familiar collective terms for discussion, giving:

Stercorarian trypanosomes:	
Chagas' disease	*T. cruzi*
Salivarian trypanosomes:	
Nagana	Principally *T. vivax, T. congolense* and *T. brucei brucei*
Sleeping sickness	*T. brucei gambiense* and *T. brucei rhodesiense*
Surra	*T. evansi* and *T. equinum*
Dourine	*T. equiperdum*

The epidemiology of these diseases, with the probable exception of dourine, involves infections in wild mammals. Epidemiological understanding, however, is usually inadequate principally because of the small size and miscellaneity of the wild animal samples examined, the inefficiency of the diagnostic methods used and the inadequacy of morphological criteria to identify pathogenic types. Immunological methods for defining and recognizing particular antigenic types, at present in process of development, may be informative in this respect in the future if applied concurrently to the wild and domestic mammals coexisting in the same areas. Biochemical studies of defined material are likely also to be informative. Such studies may lead to some immunological method of protecting domestic animals against the pathogenic effects of infection.

Finally it must be recognized that although adult wild animals are regarded as typically in non-pathogenic balance with these trypanosome infections it is not known if this situation is purchased at some cost to the population as, for example, of abortions or of neonatal or juvenile mortality.

INTRODUCTION

Trypanosome infections are known to be widely distributed among mammals both taxonomically and geographically. Infections appear typically benign but there exists a small number of trypanosome species which appear pathogenic to some hosts which, in the main, are not usually involved in transmission cycles. When these hosts happen to be man or his domestic animals, very important disease situations may be provoked. This paper outlines our knowledge of the epidemiology of these diseases relating to wild mammalian hosts and suggests how additional information may be acquired.

Mammalian hosts are referred to by the nomenclature of Morris (1965).

CLASSIFICATION OF MAMMALIAN TRYPANOSOMES

Wenyon (1926) used the character of the life cycle to divide the genus *Trypanosoma* into those species developing in the mid and hind gut of insect vectors (metacyclic forms in the insect faeces; contaminative transmission) and those developing in the mid-gut and mouth parts or mouth parts alone of insect vectors (metacyclic forms in the insect saliva; inoculative transmission). Hoare classified mammalian trypanosomes on this basis, offered modifications in several papers, introduced the terms Stercoraria and Salivaria to name the posterior and anterior developing trypanosomes, respectively, and divided both into four subgenera (Table I).

TABLE I

Classification of the members of the genus Trypanosoma infecting mammals (simplified from Hoare, 1966)

Section	Subgenus	Type species	Other selected species
Stercoraria	Megatrypanum	T. theileri	T. tragelaphi, T. ingens
	Herpetosoma	T. lewisi	T. duttoni, T. rangeli
	Schizotrypanum	T. cruzi	T. vespertilionis
	Endotrypanum	T. schaudinni	—
Salivaria	Duttonella	T. vivax	T. uniforme
	Nannomonas	T. congolense	T. dimorphon, T. simiae
	Pycnomonas	T. suis	
	Trypanozoon	T. brucei	T. evansi, T. equinum, T. equiperdum*

*T. gambiense and T. rhodesiense are omitted following the suggestion that they are synonymous with T. brucei (Ormerod, 1967).

Anomalies and difficulties are evident. *T. rangeli*, for example, included with the Stercoraria, is now known to have an anterior development in reduviid bugs (Tobie, 1964; D'Alessandro, 1964) and of the Salivaria, *T. evansi* and *T. equiperdum* are not known to have cyclical development in an insect vector. Further development of Hoare's (1966) classification is likely when more is known of the affinities of mammalian trypanosomes to other members of the family Trypanosomatidae, in particular to those of other vertebrates.

THE TRYPANOSOME DISEASES

All the pathogenic species of trypanosomes are confined to the Salivaria with the exception of *T. cruzi* which is a member of the stercorarian subgenus *Schizotrypanum*. Description of the trypanosome diseases can be obscured by the synonymy of some of the trypanosome species involved and by the many vernacular names for the diseases they produce. Grouping of trypanosomes under selected names of diseases can provide a basis for discussion (Table II). "Chagas' disease" is the reduviid bug transmitted disease of man in South and Central America; "sleeping sickness" refers to the tsetse-fly (*Glossina*) transmitted trypanosomes of domestic animals in Africa; "surra" to the *Trypanosoma evansi* and *T. equinum* infections of domestic animals which between them, can be found in South and Central America, North Africa and enormous areas of Asia and "dourine" to the trypanosome disease of horses transmitted at coitus. The epidemiology of all of these diseases, with the probable exception of dourine, involves infections in wild mammals.

TABLE II

An epidemiological grouping of the trypanosomes pathogenic to man or his domestic animals into trypanosome diseases

The disease	Section	Trypanosomes
Chagas' disease	Stercoraria	*T. cruzi*
Nagana	Salivaria	*T. vivax, T. uniforme,*
		T. congolense, T. dimorphon,
		*T. simiae, T. suis, T. brucei brucei**
Sleeping sickness	Salivaria	*T. brucei gambiense,**
		*T. brucei rhodesiense**
Surra	Salivaria	*T. evansi, T. equinum*
Dourine	Salivaria	*T. equiperdum*

*After Ormerod (1967).

EPIDEMIOLOGY

Chagas' disease

T. cruzi has a natural transmission cycle by bugs of the family Reduviidae (Hemiptera, Heteroptera) in a wide range of wild mammals. These include species representing *Dasypus* (armadillo), *Didelphis* (opossum), *Neotoma* (wood rat), *Procyon* (raccoon), *Pseudolopex* (fox) and *Mustela* (ferret) (Wood and Wood, 1941, 1961). The elusive habits of these animals enables them to live close to man. The involvement of dogs and cats as reservoir hosts allows the more easy infection of reduviid bugs infesting human habitation. Commonly associated with *T. cruzi* is a non-pathogenic trypanosome of man, *T. rangeli*, distinguishable in peripheral blood by its greater size, and transmitted by the same vectors. Trypanosomes morphologically similar to *T. cruzi* are widely distributed in nature particularly amongst microchiropteran (insectivorous) bats (e.g. Marinkelle, 1966).

Nagana

The trypanosomes involved in the collective term "nagana" occur in Africa within the distribution of *Glossina* species, that is, between 14°N and 29°S latitude. The trypanosomes are maintained by sylvatic cycles between *Glossina* species and wild mammals. The tsetse inhabiting savannah country, principally *G. morsitans* and *G. pallidipes*, are the most important as they can completely deny vast areas of Africa to the use of cattle. The extraordinary circumstance arises that nagana is seen in cattle at the margins of the distribution of savannah tsetse where tsetse and cattle come into occasional contact. Maps prepared by some African countries, e.g. Tanganyika (1965), show the cattle and savannah tsetse distributions as being mutually exclusive.

Despite the importance to man of these infections of wild mammals the total data can be considered meagre. Ashcroft (1959) tabulated the results of surveys in Africa to that date and a few more can now be added, e.g. Dias and Gonçalves (1963), Burridge, Pullan, Reid, Sutherst and Wain (1966), Baker, Sachs and Laufer (1967) and Geigy, Kauffmann and Beglinger (1967). Lumsden (1962) commented on the miscellaneity of the data up to that year but drew attention to the fact that incidences recorded for the Giraffidae and tribes Reduncini (reedbuck) and Strepsicerotini (kudu) of the Bovidae were high, in the order of 35% in comparison to other groups none of which exceeded 16%.

Ashcroft, Burtt and Fairbairn (1959) studied the susceptibility of wild mammals to *T. brucei* infections and found wide variations between species under experimental conditions. At one extreme some species were usually killed by the infection; some species survived with variations in the intensity and duration of the parasitaemia and at the other extreme the baboon (*Papio*) was found to be non-infectible. Similar studies of host range for other trypanosome species concerned do not appear to exist. Both this kind of study, however, and the data from field surveys cannot be used as a direct indication of the relative importance of mammalian species as reservoir hosts. The importance of mammalian species as a food source for *Glossina* is not related to the characteristics of the peripheral parasitaemias of salivarian trypanosomes. A susceptible mammalian species may even not normally enter *Glossina* habitat. McCulloch (1967) reported encountering two sick zebra (*Equus burchelli*) in an area adjacent to the Serengeti National Park, Tanzania, and within a distribution of *Glossina swynnertoni*, *G. pallidipes* and *G. brevipalpis*. Clinical, post-mortem, and microscopical observations lead to the conclusion that the abnormalities observed in the zebra were probably due to infection with *Trypanosoma brucei*. Zebra are not favoured hosts of tsetse (Glasgow, 1963) and the likely explanation of the episode is of an accidental involvement in natural transmission cycles.

A hypothesis can be made that wild mammalian hosts of tsetse fly suffer from trypanosome disease and the survivors acquire some immunity. Supportive evidence is lacking and may be very difficult to notice in field situations, particularly if the loss is expressed in abortions or neonatal or calfhood mortality.

Important information on wild animal infections has also come from the serological examination of blood meals of *Glossina* (Weitz, 1963; Cunningham, Kimber and Bailey, 1966). Weitz divided the 15 *Glossina* species studied into 5 groups depending on whether they feed mainly on suids, suids or bovids, mainly on bovids, mainly on mammals other than suids or bovids, or on a variety of hosts including man. These feeding patterns are not simply determined by the relative abundance of mammalian species in an area unless the variety has become very limited. It is not known however whether the tsetse are exercising a choice or whether there is some other cause, for example, coincidence in the habits of the tsetse and its "favoured host" (Glasgow, 1963). Intensive study in a single epidemiological situation may reveal which tsetse species is feeding on both wild and domestic animals and which wild animals are likely to be important reservoirs of infection. A development of the technique available for exploitation is the detection of antibodies to trypanosomes in blood meals (Cunningham, Harley, Southon and Lumsden, 1962; Cunningham *et al.*, 1966) but has only been used in relation to *T. brucei*.

Sleeping sickness

The sleeping sickness trypanosomes (*T. gambiense* and *T. rhodesiense* of Hoare's classifications) are morphologically indistinguishable from *T. brucei*. They have been separated among themselves by their infectivity and degree of pathogenicity to man and laboratory animals, their reaction to arsenicals and their transmission patterns.

T. gambiense has been considered to be independent of sylvatic cycles and to be transmitted by the riverine (*palpalis*) group of *Glossina*; *T. rhodesiense*, on the other hand, has been considered to be maintained by cycles involving the savannah (*morsitans*) group of *Glossina* and wild mammalian reservoir hosts although considerable circumstantial evidence has had one positive field corroboration from a bushbuck (*Tragelaphus*) (Heisch, MacMahon and Manson-Bahr, 1958). The close association of the habitat of this species with man and his cultivation could be very significant. Interesting also in this context was the discovery by Onyango, van Hoeve and de Raadt (1966) of an ox infected with *T. rhodesiense*.

Differentiation on epidemiological grounds has become increasingly difficult with the recovery of *T. rhodesiense* from a wild *Glossina palpalis* (Southon and Robertson, 1961) with the same tsetse species later being found responsible for an epidemic of *T. rhodesiense* (Willett, 1965). For this and other reasons Ormerod (1967) has argued that *T. gambiense* and *T. rhodesiense* should be made synonymous with *T. brucei* and be classified at a subtaxonomic level as "strains" or "demes". The suggestion allows greater freedom of discussion and is more compatible with work defining antigenic types within the same trypanosome species.

Surra

T. evansi and *T. equinum* are morphologically indistinguishable excepting that in the latter species Romanowsky stains are unable to reveal the kinetoplast (Hoare, 1950) due to the lack of DNA in its structure (Mühlpfordt, 1963). Epidemiologically also the two trypanosomes do not appear to differ and so can be considered together.

Outbreaks of surra are normally related to seasonal peak incidence of biting flies particularly Tabanidae, some of which can be shown in the laboratory to be capable of the mechanical transmission of trypanosomes. Because of this, natural transmission is assumed to be non-cyclical. Cattle, buffalo and pigs are capable of being reservoir hosts and it seems possible that cycles of transmission can occur involving domestic animals alone.

Information on infections of free-living wild animals is fragmentary and appears to relate to disease episodes, probably indicating that the mammals concerned were not normally involved in transmission cycles. Records of natural infections of *T. evansi* include *Elephas maximus* (Indian elephant) (Evans, 1910) and *Cervus unicolor* (Adams and Lionnet, 1933) and for *T. equinum, Hydrochoerus hydrochaeris* (Capybara) (Migone, 1910). Experimental infections reveal wide possibilities. These have been described for *T. evansi* in Cervidae (deer), Cercopithecidae (old world monkeys), Sciuridae (squirrels) and Cricetidae (deer mice) (Packchanian, 1938; Kraneveld and Mansjoer, 1952) and for *T. equinum* in *Didelphis azarae* (opossum) and *Pseudolopex* (fox) (Fornari Boehringer and de Boahringer, 1964).

Dourine

Experimentally *T. equiperdum* will infect Cercopithecidae (old world monkeys) and many of the Rodentia (Mesnil and Rouget, 1907; Packchanian, 1963) but there is no evidence of natural wild mammalian infections.

DISCUSSION

Review of the available information reveals surprising inadequacies in the understanding of the epidemiological status of wild mammalian hosts. Much information still dates from the beginning of the century when trypanosomiasis was a new excitement. The inadequacies appear to lie in the small size and miscellaneous nature of the wild mammalian samples examined, the inefficiency of the diagnostic methods used and the difficulties of using morphological criteria to identify species.

Methods of diagnosis may be either parasitological or serological. Parasitological methods either involve the direct examination of a tissue or the attempt to multiply trypanosomes which may be present. This may be achieved by the inoculation of tissue into a susceptible laboratory animal, culture, or the feeding of laboratory-bred vectors on suspected hosts (xenodiagnosis). Serological diagnosis has had limited success. A complement fixation test enabled the elimination of dourine from Canada and a similar test can be used with effect to diagnose human infections of *T. cruzi* (Woody, Hernandez and Suchow, 1965). Bailey, Cunningham and Kimber (1967) developed an indirect flourescent antibody technique applied to dried blood for use as a screening test in the diagnosis of sleeping sickness which has probable application to the nagana trypanosomes. In addition, the detection of immunologically induced changes in serum proteins can be made by crude precipitation (Leach, 1961) or by refined protein recognition techniques (Lumsden, 1967).

Difficulties of diagnosis may operate least for Chagas' disease. Interpretation of results in wild mammalian infections cannot, however, be made with certainty without knowledge of the infectivity of an isolation to man. A similar difficulty exists with sleeping sickness where the requirement of inoculating human volunteers with *brucei*-like organisms derived from mammalian hosts to determine their infectivity to man is an obvious limiting factor. Sufficient is known of the nagana trypanosomes to at least provide epidemiological outlines. The epidemiological status of specific mammalian hosts may however only be conjectured and not defined. The two most important trypanosomes *T. congolense* and *T. vivax* are not reliably diagnosed in chronic infections of susceptible domestic animals and possibly less so in wild mammalian hosts. The adequacy of available diagnostic methods varies with trypanosome species and the comparative incidence in a population of mammalian hosts has to be guessed. Xenodiagnosis does not help; the infection of tsetse is not a practical laboratory procedure because infection rates are low. The greatest lack of knowledge is evident with surra in relation to which surveys of wild animals do not appear to have been attempted.

The unreliability of methods available for the diagnosis of trypanosome infections with submicroscopic parasitaemia is the heart of the matter. The lack of application of serological techniques is due in large part to the difficulties imposed by the antigenic lability of trypanosome populations. The same characteristic has prevented the development of vaccination procedures. Antigenic variants are also important epidemiologically because other characteristics may vary, such as resistance to trypanocidal drugs, infectivity to differing hosts and pathogenicity. The patterns of antigenic variation in a trypanosome population and their distribution can therefore be argued as a basic requirement and may incidentally resolve problems of synonymy.

In Nigeria, Gray (1966) has already attempted the use of immunological methods to study aspects of the epidemiology of *T. brucei*. He found, for example, that isolates of *T. brucei* from cattle in widely separated parts of Nigeria had few antigens in common but close antigenic relationships could be demonstrated in isolates from different animals in the same herd. Extension of this work not only to other salivarian trypanosome species but to isolates from both the domestic and wild mammals sharing the same environment is likely to give considerable information pertinent to the practicability of vaccination as a means of control. Exploitation of these studies must depend on a recognized nomenclature for antigenic variants, as is common practice in other areas of microbiology, enabling the easy exchange of antigen material between interested workers. If it is discovered that populations of particular biological behaviour cannot be identified on antigenic characters alone, biochemical tools may have to be developed in association. This whole line of approach can be further argued as being not only applicable to all the pathogenic trypanosome species but also to other disease complexes, e.g. the tick transmitted *Theileria* spp.

SUMMARY

An outline is given of the present knowledge of the epidemiology of the trypanosome diseases of man and his domestic animals in relation to trypanosome infections of free-living mammalian hosts. Reasons are offered for important gaps in knowledge and a serological approach is suggested as the most likely to achieve greater understanding.

REFERENCES

Adams, A. R. D. and Lionnet, F. E. (1933). An outbreak of Surra among the wild deer (*Cervus unicolor var.*) of Mauritius. *J. comp. Path. Ther.* **46**, 165–167.

Ashcroft, M. T. (1959). The importance of African wild animals as reservoirs of trypanosomiasis. *E. Afr. med. J.* **36**, 289–297.

Ashcroft, M. T., Burtt, E. and Fairbairn, H. (1959). The experimental infection of some African wild animals with *Trypanosoma rhodesiense*, *T. brucei* and *T. congolense*. *Ann. trop. Med. Parasit.* **53**, 147-161.

Bailey, N. M., Cunningham, M. P. and Kimber, C. D., (1967). The indirect fluorescent antibody technique applied to dried blood, for use as a screening test in the diagnosis of human trypanosomiasis in Africa. *Trans. R. Soc. trop. Med. Hyg.*, **61**, 696–700.

Baker, J. R., Sachs, R. and Laufer, I. (1967). Trypanosomes of wild mammals in an area north-west of the Serengeti National Park, Tanzania. *Z. Tropenmed. Parasit.* **18**, 280–284.

Burridge, M. J., Pullan, N. B., Reid, H. W., Sutherst, R. W. and Wain, E. B. (1966). Survey of trypanosomiasis in wild animals in Busoga District, Uganda. *XI Int. sci. Counc. Tryp. Res.* 51–52.

Cunningham, M. P., Harley, J. M. B., Southon, H. A. W. and Lumsden, W. H. R. (1962). Detection of antibodies in blood meals of haematophagons Diptera. *Science, N.Y.* **138**, 32–33.

Cunningham, M. P., Kimber, C. D. and Bailey, N. M., (1966). Identification of the blood meals of blood sucking insects. *XI Int. sci. Counc. Tryp. Res.* 121–122.

D'Alessandro, A. (1964). New aspects of the life cycle of *Trypanosoma rangeli*. *Int. Congr. trop. Med. Malar.* No. 7. **2**, 258–259.

Dias, T. S. and Gonçalves, C. B. (1963). Trypanosoma infection in game and tsetse flies in the Muda region of Mozambique. *Anais Servs Vet. Ind. anim. Moçamb.* No. 9, 41–53.

Evans, G. H., (1910). Elephant surra. Trypanosomiasis in the elephant. A preliminary note. *J. trop. vet. Sci.* **5**, 233–239.

Fornari, D. E., Boehringer, E. G. and de Boahringer, I. K. (1964). Sensitivity of wild animals (opossum and fox) to *Trypanosoma equinum*. *Revta Investnes Agropec*. Ser. 4. **1**, 59–65. (Abstract in *Vet. Bull*. 1965, **35**, 766).

Geigy, R., Kauffmann, M. and Beglinger, R. (1967). A survey of wild animals as potential reservoirs of trypanosomiasis in the Ulanga District (Tanzania). *Acta. trop*. **24**, 97–108.

Glasgow, J. P. (1963). *The distribution and abundance of tsetse*. Pergamon Press, Oxford.

Gray, A. R. (1966). Immunological studies on the epizootiology of *Trypanosoma brucei* in Nigeria. *XI Int. sci. Counc. Tryp. Res*. 57–61.

Heisch, R. B., MacMahon, J. P. and Manson-Bahr, P. E. C. (1958). The isolation of *Trypanosoma rhodesiense* from a bushbuck. *Br. med. J*. **2**, 1203–1204.

Hoare, C. A. (1950). Akinetoplastic strains of *Trypanosoma evansi* and the status of allied trypanosomes in America. *Revta. Soc. mex. Hist. nat*. **10**, 81.

Hoare, C. A. (1966). The classification of mammalian trypanosomes. *Ergebn. Mikrobiol. ImmunForsch. exp. Ther*. **39**, 43–57.

Kraneveld, F. C. and Mansjoer, M. (1952). Onderzoekingen over de geroeligheid voor surra II Het verloop der ziekte bij enkole in het wild levelde dieren in Indonesie. *Hemera Zoa* **59**, 117–146.

Leach, T. M. (1961). Observations on the treatment of *Trypanosoma evansi* infection in camels. *J. comp. Path. Ther*. **71**, 109–117.

Lumsden, W. H. R. (1962). Trypanosomiasis in African Wildlife. *Int. Conf. Wildl. Disease* No. 1, 66–68.

Lumsden, W. H. R. (1967). Trends in research on the immunology of trypanosomiasis. *Bull. Wld Hlth Org*. **37**, 167–175.

McCulloch, B. (1967). Trypanosomes of the *brucei* subgroup as a probable cause of disease in wild zebra (*Equus burchelli*) *Ann. trop. Med. Parasit*. **61**, 261–264.

Marinkelle, C. J. (1966). Observations on human, monkey and bat trypanosomes and their vectors in Colombia. *Trans. R. Soc. trop. Med. Hyg*. **60**, 109–116.

Mesnil, F. and Rouget, I. (1907). Sensibility of ruminants and monkeys to the trypanosome of dourine. *J. trop. vet. Med*. **2**, 177–183.

Migone, L. E. (1910). Le role des carpinchos comme reservoir de virus dans la conservation du mal de caderas. *Bull. Soc. Path. exot*. **3**, 524.

Morris, D. (1965). *The Mammals*. Hodder and Stoughton, London.

Mühlpfordt, H. (1963). Über die bedeutung und feinstruktur des blepharoplasten bei parasitischen Flagellaten I, II. *Z. Tropenmed. Parasit*. **14**, 357–475.

Onyango, R. J., van Hoeve, K. and de Raadt, P. (1966). The epidemiology of *Trypanosoma rhodesiense* sleeping sickness in Alego location, Central Nyanza, Kenya. I. Evidence that cattle may act as reservoir hosts of trypanosomes infective to man. *Trans. R. Soc. trop. Med. Hyg*. **60**, 175–182.

Ormerod, W. E. (1967). Taxonomy of the sleeping sickness trypanosomes. *J. Parasit*. **53**, 824–830.

Packchanian, A. (1938). Susceptibility and resistance of various species of *Peromyscus* (American deer mice) to infection with *Trypanosoma hippicum* and the possibility of certain "wild mice" being reservoir hosts to pathogenic trypanosomes. *Am. J. trop. Med*. **18**, 587–593.

Packchanian, A. (1963). Susceptibility and resistance of American rodents to *Trypanosoma equiperdum*. *Tex. Rep. Biol. Med*. **21**, 380–390.

Southon, H. A. W. and Robertson, D. H. H. (1961). Isolation of *Trypanosoma rhodesiense* from wild *Glossina palpalis*. *Nature, Lond*. **189**, 411.

Tanganyika (1956). *Atlas of Tanganyika* Dar es Salaam: Government Press.

Tobie, E. J. (1964). Increased infectivity of a cyclically maintained strain of *Trypanosoma rangeli* to *Rhodnius prolixus* and mode of transmission by invertebrate host. *J. Parasit.* **50**, 593–598.

Weitz, B. (1963). The feeding habits of *Glossina. Bull. Wld Hlth Org.* **28**, 711–729.

Wenyon, C. M. (1926). *Protozoology.* Bailliére, Tindall and Cox, London.

Willett, K. C. (1965). Some observations on the recent epidemiology of sleeping sickness in Nyanza region Kenya, and its relation to the general epidemiology of gambian and rhodesian sleeping sickness in Africa. *Trans. R. Soc. trop. Med. Hyg.* **59**, 374–394.

Wood, F. D. and Wood, S. F. (1941). Present knowledge of the distribution of *Trypanosoma cruzi* in reservoir animals and vectors. *Am. J. trop. Med.* **21**, 335–345.

Wood, S. F. and Wood, F. D. (1961). Observations on vectors of "Chagas' disease" in the United States III New Mexico. *Am. J. trop. Med. Hyg.* **10**, 155.

Woody, N. C., Hernandez, A. and Suchow, B. (1965). American trypanosomiasis III. The incidence of serologically diagnosed Chagas' disease among persons bitten by the insect vector. *J. Pediat.* **66**, 107–109.

H

Symp. zool. Soc. Lond. (1968) No. 24, 147–158.

TRYPANOSOMES OF WILD MAMMALS IN THE NEIGHBOURHOOD OF THE SERENGETI NATIONAL PARK

J. R. BAKER

Department of Parasitology, London School of Hygiene and Tropical Medicine, Keppel Street, London, England

SYNOPSIS

The paper reviews recent reports on the incidence of trypanosome infections in wild mammals in or near Serengeti National Park, Tanzania. One hundred and twenty-nine such mammals, belonging to 17 species, have been examined by various workers and 42 (33%) were found to be infected with salivarian trypanosomes: the results are given in detail.

The high infection rates in carnivores (lion and hyaena) and in herbivores not normally fed on by tsetse flies (*Glossina*) lead to the suggestion that non-cyclical transmission of trypanosomes by blood-sucking flies other than *Glossina* is playing an important rôle in maintaining the infection in this area; it is further suggested that carnivores become infected by feeding on infected herbivores.

Reports of pathogenic infections with salivarian trypanosomes in wild mammals are rare. Two (in zebra) have been reported from this area, with another more doubtful example (in a lion cub).

INTRODUCTION

The incidence of trypanosome infections in the wild mammals of the Musoma district, Mara region, Tanzania was fairly intensively studied in 1966 and 1967, largely due to an apparent recrudescence of human trypanosomiasis in the area (which includes the Serengeti National Park) under conditions which suggested the presence of a mammalian reservoir host of the disease (see Baker, Sachs and Laufer, 1967). The purpose of this paper is to review and discuss the results obtained in relation to the wild game animals of the area.

PREVIOUS WORK

Several surveys of trypanosome infections in wild mammals have been made in different parts of East Africa: the results of the more important of these are summarized by Ashcroft (1959) and references given by him will not be repeated here. The most frequently infected animals were waterbuck,* kudu, reedbuck, giraffe, bushbuck and eland, all of which had infection rates greater than 25%. Thomson's gazelle, of which only 10 were examined, were never found to be infected and wildebeest only rarely so (2·8%). Taking into consideration the infection rates in the different species of mammal and their importance as a source of food for the prevalent tsetse flies (*Glossina* spp.), Ashcroft concluded that the warthog was the most important reservoir host of trypanosomiasis of game animals in the areas covered by the surveys reviewed in his paper.

A more recent survey, conducted by Geigy, Kauffmann and Beglinger (1967) in southern Tanzania, found that reedbuck were the most frequently infected of the 8 species examined.

METHODS

The methods used in diagnosing trypanosome infections and in identifying the species concerned are given in the papers cited below. Generally, thick and thin blood films were prepared from each animal examined and blood was inoculated into one or two white rats according to the techniques described by Baker *et al.* (1967). The taxonomic nomenclature of the trypanosomes follows that of Hoare (1966).

RESULTS

The results of the recent surveys conducted in this area are summarized in Table I. It will be seen that a total of 129 mammals, of 17 different species, has been examined although some species were represented by rather small numbers. Of these mammals, 42 were infected with one or more species of trypanosome—an overall infection rate of about 33%. Only buffalo, leopard and jackal were free from infection and the numbers of these animals examined were very small.

All strains of the subgenus *Trypanozoon* (i.e., the "*T. brucei* group") were identified as *Trypanosoma brucei* (using this name to include the subspecies† *T. brucei brucei* and *T. brucei rhodesiense*) and all such strains inoculated into rats were infective to the rodents. All but one of the strains of *Nannomonas* (the "*T. congolense* group") were identified as *T. congolense*, the exception being one strain from a warthog which might have been *T. simiae*: 13 of these 20 strains were infective to rats.

*Scientific names of wild mammals mentioned in the text are given in Tables I and II, following the usage of Allen (1939).
†Variously regarded as distinct species, subspecies or merely strains (see Ormerod, 1967).

TABLE I

Incidence of trypanosome infections in wild mammals in or near the Serengeti National Park, Tanzania

Common name	Scientific name	No. examined	No. infected with[a] Trypanozoon	Nannomonas	Duttonella	Total no. infected mammals
Buffalo	Syncerus caffer (Sparrman)	3	0	0	0	0
Eland	Taurotragus oryx (Pallas)	5	0	3	2	4
Giraffe	Giraffa camelopardalis (L.)	1	0	0	1	1
Grant's Gazelle	Gazella granti Brocke	3	0	1	0	1
Hartebeest	Alcelaphus buselaphus (Pallas)	10	1	0	1	2
Hunting Dog	Lycaon pictus (Temminck)	2[b]	0	1	0	1
Hyaena	Crocuta crocuta (Erxleben)	3[b]	0	3	0	3
Impala	Aepyceros melampus (Lichtenstein)	6	1	2	0	2
Jackal	Thos mesomelas (Schreber)	2[b]	0	0	0	0
Leopard	Felis pardus L.	1	0	0	0	0
Lion	F. leo Neumann	11[c]	7	2	0	9
Thomson's Gazelle	Gazella thomsonii Gunther	7	0	1	0	1
Topi	Damaliscus korrigum (Ogilby)	11	2	1	1	3
Warthog	Phacochoerus aethiopicus (Pallas)	14	1	2	0	3
Waterbuck	Kobus defassa (Rüppell)	6	3	1	0	4
Wildebeest	Connochaetes taurinus (Burchell)	22	4	3	1	6
Zebra	Equus burchelli (Grey)	22[d]	2	0	0	2
		129	21	20	6	42

[a] All records from Baker et al. (1967) except as indicated.
[b] Additional records (1, 2 and 1, respectively) from R. Sachs (personal communication).
[c] Two additional records from Sachs et al. (1967) and nine from R. Sachs and G. B. Schaller (personal communication).
[d] Both infected zebra reported by McCulloch (1967): the uninfected animals include 15 examined by blood films only (R. Sachs, personal communication).

TABLE II

Scientific names of wild mammals mentioned in the text and not given in Table I

Common name	Scientific name
Bush Baby	*Galago crassicaudatus* E. Geoffroy
Bushbuck	*Tragelaphus scriptus* (Pallas)
Hippopotamus	*Hippopotamus amphibius* L.
Kudu	*Strepsiceros* sp.
Reedbuck	*Redunca* sp.
Sable Antelope	*Hippotragus niger* (Harris)

Five of the 6 strains of *Duttonella* (the "*T. vivax* group") were identified as *T. vivax*; that from the giraffe was identified as *T. uniforme* because the mean length of the organisms was only $16 \cdot 6\mu$ (range $15–20\mu$; 10 individuals measured). None of the strains of this subgenus infected rats.

Parasitaemias in the infected wild mammals varied from moderate to very light; in many animals, particularly those infected with *T. brucei*, parasites were not detected in thick blood films.

Five mammals were found to be infected with two species of trypanosome: a wildebeest and an impala had *T. brucei* and *T. congolense* and another wildebeest, an eland and a topi had *T. congolense* and *T. vivax*. In addition to the results recorded in Table I, Dr. Sachs has informed me (personal communication) that the examination of thick blood films obtained by himself and Dr. Schaller from about 60 further lions has revealed trypanosome infections in some two-thirds of them: specific identification of the organisms has not yet been completed.

DISCUSSION

Incidence of infection in different species of mammal

Reference to Table I shows that four species of mammal had infection rates greater than 50%: these were hyaena (100%), lion (83%), eland (80%) and waterbuck (67%). Infection rates below 20% were found in Thomson's gazelle (14%), zebra (9%), and buffalo, jackal and leopard (zero): the number of infected zebra might have been higher had not the majority of specimens been examined only by means of blood slides and not by the inoculation of blood to rats. The numbers of buffalo, jackal and leopard examined were too low for the result to be significant.

Earlier surveys of this kind in East Africa, summarized by Ashcroft (1959), have recorded a similarly high infection rate in waterbuck (52%) but differ from the recent work in having revealed only very low infection rates in wildebeest (less than 3%, in contrast to the 27% reported in this paper). Reports of the examination of lions for trypanosome infection have hitherto been very sparse: Weck (1914) recorded the examination of 2 lions, one of which was infected with an unidentified species of trypanosome, and Baker (1960) reported another infection due to *T. congolense*. Recently Mr. J. Ford (personal communication) drew my attention to a newspaper report of two other, pathogenic, infections ascribed to *T. brucei* (Adamson, 1965). The possible significance of some of these results will be discussed in the subsections below.

The record of *T. uniforme* from the giraffe is interesting as previously this parasite has been recorded mainly from the western part of Uganda and the eastern part of the ex-Belgian Congo and only rarely from Zululand and Tanzania (R. Killick-Kendrick, unpublished review).

Transmission of trypanosomes among wild mammals

It is of course well known that trypanosomes of the subgenera studied in the work reported in this paper (the section Salivaria of the classification introduced by Hoare, 1966) are normally transmitted by tsetse flies of the genus *Glossina* (Diptera, Cyclorrhapha), in which they undergo a cyclical development ending with the production of infective ("metacyclic") organisms in the salivary glands or mouthparts of the flies. Non-cyclical (or "mechanical") transmission by blood-sucking Diptera of other families is also known to occur. When an insect which is feeding on an infected mammal is disturbed and flies to another nearby host to complete its meal, trypanosomes still present in its proboscis, in the blood of the first host, may then be introduced into the blood-stream of the second host without having undergone any development in the insect.

Tsetse flies are common in and around the Serengeti National Park: in the area where most of the work recorded in this paper was done, the commonest species were *G. swynnertoni* and *G. pallidipes*, with *G. brevipalpis* present sporadically. The favourite hosts of the two former species are warthog and bushbuck, respectively, with warthog serving as a substitute host for *G. pallidipes* in areas from which bushbuck are absent (Weitz, 1963). *G. brevipalpis* feeds mainly on bushpig and hippopotamus. In the area under consideration, bushbuck are rarely seen: thus it would seem likely that the animal most frequently bitten by tsetse flies is the warthog. It is therefore surprising that this animal was amongst those with the lower trypanosome infection rates (21%). Also surprisingly, the animals with the highest infection rates of all, with the exception of the eland, constitute less than 1% of the food of these species of *Glossina* (eland form 1·4% of the food of *G. swynnertoni*). Furthermore, of the 4 species of mammal which are "definitely ignored by all tsetse flies as a source of food, except on isolated occasions. . ." (Weitz, 1963), 3 (waterbuck, impala and wildebeest) had fairly high infection rates (67%, 33% and 27%, respectively), while the fourth (zebra) was more rarely infected (9%).

The significance of this, concerning transmission, is thought to be two-fold (its possible relationship to pathogenicity will be discussed below). Firstly, the high proportion of infected herbivores of species which are rarely if ever bitten by tsetse flies must indicate a considerable amount of non-cyclical transmission by insects such as *Tabanus* and *Stomoxys*. The density of the game animal population in the Serengeti Park would doubtless facilitate this, by making it easier for a disturbed fly to alight on another host within a sufficiently short time for its proboscis to retain the blood (and, possibly, trypanosomes) of its previous host. Also, the existence close together of herds of different species, including both those virtually never fed on by tsetse (impala, wildebeest) and those fed on more regularly (warthog, eland), probably allows the trypanosomes to gain a foothold more readily in the former groups, as a result of non-cyclical transmission from the latter.

Secondly, the remarkably high incidence of trypanosome infection amongst carnivores (over 80% of the lions and hyaenas examined were infected) suggests strongly that these animals, which are also rarely used by tsetse as a source of food, become infected by killing and eating infected herbivores. The possibility of the infection being transmitted in this way has been demonstrated by several workers. The earlier observations are reviewed by Duke, Mettam and Wallace (1934) who themselves infected cats in this fashion: more recently, Heisch (1963) infected a bush baby with *T. brucei rhodesiense* by allowing it to kill and eat an infected laboratory rat. It is probable in such instances that the trypanosomes penetrate small lesions of the buccal mucosa which must often be slightly damaged by the bones of the prey. Successful infection by this means would presumably depend on the predator's eating the prey soon after the latter had been killed: carrion feeders would be less likely to become infected.

Thus Ashcroft's (1959) calculations of the relative importance of various species as reservoirs of trypanosomiasis may lead to erroneous conclusions, as far as trypanosomiasis of mammals other than man is concerned, because they do not take account of either of the above means of transmission (non-cyclical and oral) which, recent results suggest, may account for a considerable proportion of the transmission of trypanosomes occurring among wild mammals at least in certain situations.

Pathogenicity of trypanosomes to wild mammals

Records of trypanosomes being pathogenic to wild mammals, under natural conditions, are very rare. In the work summarized in this paper, only the two zebra studied by McCulloch (1967) and one young lion examined by Sachs, Schaller and Baker (1967) were noticeably ill. In the two zebra, there is good evidence that the disease was due to the *T. brucei* infection and it is interesting that this is probably the only report of the organisms having invaded the central nervous system of a naturally infected wild mammal. It is less certain that the lion cub was ill as a result of its trypanosome infection (*T. brucei*).

The likelihood of finding animals of a certain species infected with trypanosomes depends not only on whether they become infected but also on their response to infection. Members of a highly susceptible species will probably be rapidly killed by the parasites and so the chance of finding infected individuals will be reduced. On the other hand, very tolerant animals may have such low parasitaemias that diagnosis becomes difficult, particularly when it is dependent on the finding of organisms in blood films. Although, in most of the recent work summarized here, rat inoculation was used as an aid to diagnosis of infection, *T. vivax*, *T. uniforme* and *T. simiae*, as well as some strains of *T. congolense* (see Godfrey, 1961), are unable to infect rats: hence their recognition was dependent upon sufficient intensity of parasitaemia to make detection in thick blood films possible. There is little evidence of the pathogenicity to wild mammals of species of *Trypanosoma*: such as there is applies exclusively to *T. brucei* (amongst the Salivaria). The infections of lions reported by Adamson (1965) were, apparently, pathogenic. Ashcroft (1959) has summarized the effect of experimental infection with *T. brucei* on several species of mammals: hyaena, impala and warthog were said to be "usually tolerant" to infection, while Thomson's gazelle and jackal were "usually killed" by it. Thus the apparent absence of *T. brucei* infection from Thomson's gazelle, both in the work recorded in this paper and in that summarized by Ashcroft (1959), may have been due to the virulence of this organism to the gazelle. More information on the pathogenicity of the salivarian trypanosomes to game animals, and on the pathology of the infections produced, is badly needed.

SUMMARY

One hundred and twenty-nine wild mammals, belonging to 17 species, have been examined by various workers during 1966 and 1967, in or near the Serengeti National Park, Tanzania. Forty-two (33%) of these mammals were infected with species of *Trypanosoma* belonging to the subgenera *Trypanozoon* (21 infections), *Nannomonas* (20 infections) and *Duttonella* (6 infections): the latter included one infection in a giraffe (*Giraffa camelopardalis*) with *T.* (*Duttonella*) *uniforme*. The most commonly infected mammals were hyaena (*Crocuta crocuta*), lion (*Felis leo*), eland (*Taurotragus oryx*) and waterbuck (*Kobus defassa*) with infection rates of 67-100%.

It is suggested that non-cyclical transmission of the trypanosomes, by blood-sucking insects other than *Glossina*, plays an important part in infecting herbivores in this area and that predatory carnivores probably often become infected by eating carcasses of infected herbivores.

There are very few reports of trypanosomes being pathogenic to naturally infected wild mammals. In the work summarized in this paper, two such instances occurred in zebra (*Equus burchelli*) infected with *T.* (*Trypanozoon*) *brucei* (McCulloch, 1967) and another, less certainly, in a young lion infected with the same species (Sachs *et al.*, 1967).

ACKNOWLEDGEMENTS

I am very grateful to Dr. R. Sachs and Dr. G. B. Schaller for allowing me to quote their unpublished results.

REFERENCES

Adamson, G. (1965). Learning to live free. *Times*, London, 7 July 1965, p. 12.
Allen, G. M. (1939). *A checklist of African Mammals.* Museum of Comparative Zoology, Harvard, Cambridge, Massachusetts.
Ashcroft, M. T. (1959). The importance of African wild animals as reservoirs of trypanosomiasis. *E. Afr. med. J.* **36**, 289-297.
Baker, J. R. (1960). A trypanosome of *T. congolense* group in African lion and leopard. *Trans. R. Soc. trop. Med. Hyg.* **54**, 2.
Baker, J. R., Sachs, R. and Laufer, I. (1967). Trypanosomes of wild mammals in an area northwest of the Serengeti National Park, Tanzania. *Z. Tropenmed. Parasit.* **18**, 280-284.
Duke, H. L., Mettam, R. W. M. and Wallace, J. M. (1934). Observations on the direct passage from vertebrate to vertebrate of recently isolated strains of *Trypanosoma brucei* and *Trypanosoma rhodesiense. Trans. R. Soc. trop. Med. Hyg.* **28**, 77-84.

Geigy, R,. Kauffmann, M. and Beglinger, R. (1967). A survey of wild animals as potential reservoirs of trypanosomiasis in the Ulanga District (Tanzania). *Acta trop.* **24**, 97–108.

Godfrey, D. G. (1961). Types of *Trypanosoma congolense*. II. Differences in the course of infection. *Ann. trop. Med. Parasit.* **55**, 154–166.

Heisch, R. B. (1963). Presence of trypanosomes in bush babies after eating infected rats. *Nature, Lond.* **169**, 118.

Hoare, C. A. (1966). The classification of mammalian trypanosomes. *Ergebn. Mikrobiol. ImmunForsch. exp. Ther.* **39**, 43–57.

McCulloch, B. (1967). Trypanosomes of the *brucei* subgroup as a probable cause of disease in wild zebra (*Equus burchelli*). *Ann. trop. Med. Parasit.* **61**, 261–264.

Ormerod, W. E. (1967). Taxonomy of the sleeping sickness trypanosomes. *J. Parasit.* **53**, 824–830.

Sachs, R., Schaller, G. B. and Baker, J. R. (1967). Isolation of trypanosomes of the *T. brucei* group from lion. *Acta trop.* **24**, 109–112.

Weck, (1914). Beobachtungen über Trypanosomen des Menschen und der Tiere am Rovuma-flusse. *Arch. Schiffs- u. Tropenhyg.* **18**, 113–124.

Weitz, B. (1963). The feeding of *Glossina*. *Bull. Wld Hlth Org.* **28**, 711–729.

DISCUSSION

LUMSDEN: The information gathered together by Ashcroft depended on records, from many areas, which dated back to the beginning of this century and involved many different standards of observation. Whilst this kind of information gave some picture of the parasitaemia in wild animals, it might not be an entirely reliable guide to the epidemiological importance of an animal species. Such incidence is likely to be determined by the reaction of the host as well as the frequency of infection: some other method is required, such as the feeding on hosts of groups of a standard vector arthropod to determine reliably the significance of a species as a source of infection. Mechanical transmission might be overestimated as a method of infection of wild animals. The distribution of cattle is closely related to the distribution of *Glossina* and in many areas reputed to have "trypanosomiasis without tsetse", tsetse have been subsequently detected. Dr. Baker is absolutely right in suggesting that carnivores are infected by the ingestion of infected meat; the waterbuck may simply be an animal with a species characteristic to show a high parasitaemia. With regard to the Chairman's question concerning the effect of stress, there was an example of a rhinoceros which showed a high parasitaemia with trypanosomes and which died after capture.

HOARE: Mr. Wells has commented on the apparent absence of reservoir hosts of *Trypanosoma evansi* which distinguishes it from the tsetse-borne species. The clue to this difference lies in the past history of these trypanosomes. It is reasonable to assume that in tropical Africa wild mammals harboured nagana parasites long before man and his livestock had arrived on the scene and become involved. On the other hand, there is good circumstantial evidence that *T. evansi* had originated from *T. brucei* near the northern boundaries of *Glossina*, whence it was introduced through human agency (with camels) first into North Africa, then into Asia and eventually reaching the New World. As *T. evansi* was a newcomer outside Africa, there could not have been any wild mammals naturally infected with this species. Hence it follows that the infections reported from deer in Mauritius and from vampire bats in South America are secondary acquisitions resulting from the introduction of *T. evansi* into new areas.

VICKERMAN: Is anything known about the duration of trypanosome infections in game animals? This could be an important factor in determining the incidence of infection. Does an animal remain infected for life or does it manage to throw off the infection? In the laboratory, the deer mouse (*Peromyscus*) eventually throws off the infection.

BAKER: The results of the Tinde experiment might provide the answers to the question on the duration of the infection. Referring to the question of mechanical or non-cyclical transmission, I do not mean to imply that such a method would be sufficient to maintain the organisms but that, given the occurrence of tsetse transmission between certain mammalian species, non-cyclical transmission might serve, in some areas, to spread the parasites into mammal populations not normally fed upon by *Glossina*.

HOARE: As regards the duration of the infection in wild mammals, some light on this question is thrown by two cases observed here in the London Zoo: (i) a hyaena (*Crocuta crocuta*), received from Uganda in 1936, was found to harbour *T. brucei* when its blood was examined *post mortem* in 1942; (ii) an antelope (*Tragelaphus gratus*), which arrived from Nigeria in 1934 and died at the Zoo in 1945, was found to harbour *T. congolense*. In both instances it is obvious that the infections were acquired in Africa, so that their duration exceeded 6 years in the first case and 11 years in the second.

COX: Is there any direct evidence that animals such as waterbuck are *not* fed on by tsetse flies or was all the evidence based on serological analysis of tsetse blood meals? Can trypanosomes be transmitted between such animals by tsetse flies which simply probe but do not take a full blood meal?

HALL: Is there any evidence that South American primates, particularly squirrel monkeys, normally suffered ill effects from parasitaemia with *Trypanosoma cruzi* or were they solely wild reservoirs of the organism with patent disease only being activated, for example, by stress such as that imposed by captivity. My question is prompted by personal clinical and laboratory investigations which suggested the possibility that Chagas' disease was the cause of illness in a squirrel monkey colony.

WELLS: There are several records of *T. cruzi* being isolated from monkeys, e.g. by Marinkelle in Columbia; I am unaware of any records of *T. cruzi* causing symptoms in monkeys or other free-living wild animals. There are, however, records of symptoms in domestic animals, particularly dogs.

MARSDEN: In experimental infections there is quite a marked range of reactions. Using a Peru strain of *Trypanosoma cruzi* we have infected a variety of animals. A dose of trypanosomes from mouse blood, calculated on a body weight basis (usually 100 trypanosomes/10 g body weight), has been used. Rhesus monkeys develop a high parasitaemia with Romaña's sign but subsequently recover. Pigs suffer a mild infection with parasitaemia that can be demonstrated by subinoculation. Various strains of albino mice usually die but deer mice are protected; usually they do not exhibit any parasitaemia but become positive to the complement fixation test. Rabbits develop a parasitaemia but are relatively unaffected. There is, therefore, a wide variety of responses in different hosts but it must be emphasized that these are laboratory experiments. Romaña did, of course, demonstrate his sign (unilateral conjunctivitis) in experimentally infected monkeys. It is possible that a large number of wild primates in South America are infected. I do not know, at the present time, of any mammal that cannot be experimentally infected. Were Dr. Weitz's data on tsetse blood meals corrected for the actual numbers of host animals in the area? Obviously, if there were a lot of bush pigs in an area the tsetse would bite them rather than choose the occasional lion. My second question is concerned with the pathogenicity of trypanosomes in relation to their previous animal hosts. It has been noticed, for instance, that in South America cases induced by blood transfusions in man are extremely pathogenic. Armadillo infections may be fatal to man and yet it is a benign infection in the armadillo. Could we have the speaker's views on the effect of the previous host, especially in cases where there was no arthropod vector interposed?

BAKER: Dr. Weitz's results were not statistically corrected for the density of animal population but there was little doubt that the conclusions were valid. In some areas, where the favoured host species was known to form only a minority of the available mammals, a high proportion of the tsetse blood meals would still be from the favourite host in spite of its lack of availability.

GARNHAM, CHAIRMAN: In other diseases, such as plague, leishmaniasis, relapsing fever, etc., when infections are derived from animal sources, they are much less virulent to man than when man-to-man transmission occurred although this was not inevitably so.

BARNETT: (in reply to a previous question) The lioness "Elsa" was found to be infected with *Trypanosoma congolense* but she had died of piroplasmosis (*Nuttallia felis* infection).

ROTH: Research results recorded and referred to in this Symposium tended to focus attention and interest mainly on the causative agent and the relationship to the host organism. In doing this a lot of assumptions are made all the time in respect of the life history and ecology of the host animals and the dynamics of the host populations. Further progress in wildlife disease research in general, but particularly as regards trypanosomiasis, is largely dependent on the extent to which we shall be able to join parasitological and pathological studies with solid biological and ecological research into the life history of mammalian hosts, the interrelationships between host populations and environment and their behaviours. The amount of speculation and generalization in this respect is absolutely fantastic and often jeopardises the better understanding of the broader ecological issues involved in any disease manifestations.

Symp. zool. Soc. Lond. (1968) No. 24, 159–176.

SOME PIROPLASMS OF WILD MAMMALS

S. F. BARNETT

*Department of Animal Pathology, School of Veterinary Medicine,
Cambridge, England*

and

D. W. BROCKLESBY

*Agricultural Research Council, Institute for Research on Animal Diseases,
Compton, Near Newbury, Berkshire, England*

SYNOPSIS

The rigid host-specificity previously associated with the Piroplasmida is slowly being eroded as further successful transfers are made into hitherto exotic recipients. The confused taxonomy of the group is briefly discussed and conservative views are expressed.

Piroplasms are extremely common in wild mammals but, in general, virtually nothing is known about them. Only in the case of the African Buffalo is the knowledge at all extensive. The rôle of this animal as a reservoir of East Coast Fever (*Theileria parva* infection) is described.

Theileria-like intra-erythrocytic piroplasms so frequently found in African antelopes may belong to the genus *Cytauxzoon*. Species of this genus have so far been found in duiker, kudu and eland and in each case it appeared that the parasite was the cause of death.

A piroplasm was responsible for the death of the famous lioness "Elsa". Recent isolations of piroplasms from leopards into domestic cats has shown that the Felidae, like other groups, are host to at least two Babesidae.

Echinozoon hoogstraali is a parasite of the rock hyrax and is extraordinary for the presence of peri-erythrocytic filaments.

Nuttallia danii is probably the correct name for the various species described from *Meriones*. It is suggested that this organism could be a useful research tool for studies on piroplasmoses.

Protozoologists should pay more attention to acarologists, whose surveys and host-parasite lists often provide clear pointers towards further investigations.

INTRODUCTION

A few years ago we could have presented a paper about the wild mammal reservoirs of piroplasms of domestic animals with the comfortable feeling that piroplasms were essentially host-specific parasites and that domestic stock were only at hazard from piroplasms of rather closely related wild animals. However, all this is now changed and recent results, mainly in splenectomized recipients, have shown that this rigid host-specificity may be illusory. Three splenectomized human beings have been infected with *Babesia* species; two of these infections were fatal and were thought to be due to *B. divergens*, a parasite

normally found in cattle; the third case was probably due to a wild
animal parasite (Skrabalo and Deanovic, 1957; Fitzpatrick, Cotton
Kennedy, McGeown, Oreopoulos, Robertson and Soyannwo, 1968;
Braff and Condit, 1967). B. divergens was also shown to be capable of
infecting splenectomized chimpanzees and rhesus monkeys (Garnham
and Bray, 1959) and Shortt and Blackie (1965) transferred a Babesia
from a mole to a monkey; it has even been conjectured that this common
parasite of small animals (Babesia microti) may be identical with
Nuttallia equi, a parasite of horses (Arthur, 1966). Piroplasms have
also been transferred from leopards to cats, from jackals to dogs, from
zebras to horses and donkeys and from buffaloes to cattle (Brocklesby,
Dennig and Vidler, 1965; Dennig, 1966; Neitz, 1957). Other examples
exist and it is beginning to appear that some of the piroplasms may
have amazingly catholic tastes. If this is the case this paper should
really be a detailed and systematic account of all the piroplasms so
far described from free-living wild animals together with lists showing
which ticks commonly infest wild and domestic hosts. This would be,
in the main, an indigestible catalogue of rather perfunctory descriptions
so we have decided to limit this account to some piroplasms of wild
animals that we believe to be interesting and important in the hope
that further work on these neglected parasites will be undertaken.
The selection is a personal one and workers whose particular favourite
is omitted should not feel slighted.

TAXONOMY

A glib and simple definition of the piroplasms would be "Tick
transmitted intra-erythrocytic protozoa that do not produce pigment".

And where do they fit in the taxonomic scheme? This is a very awkward
question the final answer to which must await more knowledge.
Because the piroplasms seem to lack a sexual cycle the Committee on
Taxonomy and Taxonomic Problems of the Society of Protozoologists
(Honigberg, Balamuth, Bovee, Corliss, Gojdics, Hall, Kudo, Levine,
Loeblich, Weiser and Wenrich, 1964) removed them from the Class
Sporozoa, where they had rested rather uneasily against the well-
defined coccidial and malarial parasites, and placed them with the
amoebae in Superclass Sarcodina (Class Piroplasmea; Order Piroplasmida). Workers concerned with piroplasms protested rather weakly
at this but were unable really to gainsay the Committee's opinion; all
that can be said at the moment is that a sexual cycle probably does
exist, certainly the work of Riek (1964) on Babesia bigemina indicates
this, and that until the point is proved the piroplasms should either

stay in the Class Sporozoa or should be temporarily hived off into some limbo such as the "Appendage" proposed by Levine (1961b). However, if there is confusion at the level of the higher taxa, the position within the group can only be described as chaotic. A plethora of genera has been described but most of these, such as *Smithia, Rossiella, Francaiella, Gonderia* etc., have been scuttled as synonyms.

The taxonomy that we find acceptable at the moment is as follows:

Order Piroplasmida

Family Babesidae

Genera *Babesia*

Nuttallia

Echinozoon

Family Theileridae

Genera *Theileria*

Cytauxzoon

Haematoxenus

THE AFRICAN BUFFALO AND EAST COAST FEVER

An important member of the family Theileridae is *Theileria parva*, the cause of East Coast Fever of cattle. Intra-erythrocytic piroplasms of *Theileria* species all look the same and they are extremely common: for example, in East Africa, Brocklesby and Vidler (1966) found *Theileria* piroplasms in more than 60% of thin blood films prepared from 614 wild members of the Artiodactyla. Virtually nothing is known about these parasites (Neitz, 1957, summarizes the available information). Only in the case of the African Buffalo (*Syncerus caffer*) is the body of knowledge at all considerable.

Our interest in the *Theileria* species of the buffalo was stimulated by the work of Neitz (1955, 1957) and the long series of observations made by Lawrence from 1934 to 1958 and others in Southern Rhodesia; this literature has been reviewed elsewhere (Brocklesby, 1964; Brocklesby and Barnett, 1966).

As far as we know East Coast Fever originated in East Africa. It was introduced into Southern Rhodesia with a shipment of cattle from Tanganyika in the early part of this century and then spread over the country whenever suitable tick vectors existed. The veterinary authorities there, and also in South Africa where the disease arrived a little later, fought the infection with vigour but it was not until 1958 that

I

Lawrence was able to declare Southern Rhodesia to be free from the disease.

TABLE I

Bovine Theileriases in Southern Rhodesia
(compiled from Vet. Dept. records)

Year	Mortality due to East Coast Fever	Mortality due to "theileriosis"
1935	124	?
1936	120	44
1937	128	1
1938	221	?
1939	100	13
1940	428	27
1941	76	?
1942	0	?
1943	66	?
1944	25	(10 outbreaks)
1945	About 50	(19 outbreaks)
1946	41	82
1947	Not consulted	Not consulted
1948	About 2	(15 outbreaks)
1949	0	153
1950	0	About 144
1951	0	176
1952	0	86
1953	0	559
1954	4	373
1955	0	885

Table I shows how the incidence of E.C.F. waned in Southern Rhodesia as a new disease that came to be known simply as "Theileriosis" gained momentum: originally the disease was particularly associated with the presence of wild animals and large numbers of ticks and in some areas was known as "Buffalo disease". We cannot discuss the Rhodesian observations in any detail here but they were crystallized by the subsequent work of Neitz in Zululand. An area of land known as the Corridor, between two game reserves, was cleared of tsetse and opened for cattle grazing. Almost at once the cattle began to die of a disease like East Coast Fever. It was quickly apparent that it was an unusual form: mortality ceased when the herds were moved from the Corridor even although efficient tick vectors were present on the new grazing areas. It seemed that the infected cattle did not infect ticks and this was confirmed by experiment and by blood smear examination. There was virtually no production of intra-erythrocytic piroplasms and tissue smears revealed macroschizonts in smaller numbers than expected and they were also smaller in size than those of *T. parva*. Buffaloes in the area were found to be carriers of the parasite. Here was an infection maintained in buffalo but apparently not causing disease in them. Ticks which had fed on buffalo caused a fatal disease when they fed on cattle. This illness resembled East Coast Fever but differed from it in that the life cycle of the parasite was not completed in cattle, the red cells did not become infected and thus the affected cattle could not infect fresh ticks. The schizonts were slightly different in size and were much less frequent than in fatal cases of *T. parva* infection. The significant biological difference was the failure to complete the developmental cycle in cattle and this quite rightly led Neitz to designate the parasite as a new species, *T. lawrencei*. Neitz later confirmed the field observations of the Rhodesian workers when he showed that splenectomy of recovered cattle resulted in a recrudescence that was capable of infecting ticks; it was therefore assumed that recovered cattle could become carriers of *T. lawrencei* in contrast to East Coast Fever, recovery from which at that time was believed invariably to result in a sterile immunity.

After some preliminary work concerning the susceptibility of the African buffalo to our laboratory strain of *T. parva* (Barnett and Brocklesby, 1966), we began to look for *T. lawrencei* in Kenya. This was carried out by several methods one of which consisted of making collections of ticks from wild buffaloes shot on normal control work. Collections from 14 buffaloes shot in thick bush country were made; only one buffalo yielded any engorged *Rhipicephalus appendiculatus* nymphs. Thirty of these ticks were allowed to moult to adults and some of them were placed on a steer and others on a buffalo bull called Nero. This resulted in the isolation of a parasite that will, for the moment, be called "*Theileria lawrencei* (Kenya)". All transmissions were made with the tick *R. appendiculatus* and 59 cattle were infected. Twenty-five of them died, giving a mortality rate of only 42%, much less than our laboratory strain of *T. parva* which caused a death rate of 96% in similar cattle. Since Neitz reported that a good cross-immunity existed between the South African strains of *T. parva* and *T. lawrencei*, 25 of the recovered cattle were challenged with *T. parva* at various intervals after recovery and it was confirmed that infection with "*T. lawrencei* (Kenya)" did provide a strong immunity to subsequent infection with *T. parva* and that this immunity lasted for at least three years. Thin blood films from all the infected cattle were examined for piroplasms and these were counted when present: counts and measures of the size of macroschizonts and counts of the number of nuclei were made in all cases. Very marked changes occurred on passage through cattle. In 7 out of the 11 cattle of the first and second passages no piroplasms were seen at all, in 3 others they were extremely rare and in only one animal was there any noticeable production of piroplasms and this was minimal (up to 14/1000 R.B.C.); therefore, from this aspect the parasite behaved exactly like classical *T. lawrencei*.

No piroplasms could be found in 3 of the 17 cattle of the third passage and extremely low counts were recorded from a further 12 animals. However in 2 of the cattle, piroplasm production did occur to some extent (up to 25 and 72 per 1000 R.B.C.).

The tendency for piroplasm production to increase with the passage of the organism through cattle became marked in the 18 cattle of the fourth passage. In only one animal were no piroplasms seen and low counts were recorded in a further 11 cattle. In 6 of the animals piroplasm production occurred and in one of them it was copious.

In the 12 cattle of the fifth and sixth passages high counts were the rule rather than the exception and counts that were fairly low were recorded in only 3 cattle. All these remarks apply to the maximum parasitaemia seen in individual cases.

It appeared, therefore, that the ability of "*T. lawrencei* (Kenya)" to complete its life cycle in cattle increased with its passage through them. Changes also occurred with reference to tissue parasitosis; macroschizonts were infrequent in early passages but they increased from an average of 20 per 1000 lymphocytes in the first passage to an average of 296 per 1000 lymphocytes by the sixth passage. Similarly, the small schizonts, noted by Neitz and by Lawrence as a character of *T. lawrencei*, altered during passage from $3.9\,\mu$ to 5.6μ in average diameter.

Finally we must mention briefly some results with the South African strain of *T. lawrencei* that was very kindly provided by Professor Neitz. Infected ticks obtained from a carrier bovine in South Africa were placed on two cattle in Kenya. The first animal reacted and died after an extremely short febrile reaction and the second animal had a severe fever but eventually recovered. The infection was in all respects exactly as described by Neitz as *T. lawrencei*. We infected this animal with large numbers of clean *R. appendiculatus* nymphae so that many of them dropped engorged each day over a long period. After moulting to adults the ticks of each day were checked for infectivity by being fed on susceptible cattle. Most of the ticks were not infected but those that dropped engorged 17 to 19 days after the start of the fever did transmit the parasite to two cattle which both died. From these animals a serial transmission of the parasite was established and changes similar to those that have been described for "*T. lawrencei* (Kenya)" occurred. The number of schizonts, the numbers of piroplasms, the size of schizonts and the numbers of nuclei within them all increased so that the parasite became indistinguishable from *T. parva*.

We therefore conclude that it is difficult to support the validity of *T. lawrencei* as a species; the African buffalo must be regarded as a reservoir of strains of *T. parva*. The infection is maintained in wild buffaloes but it is probable that they originally acquired *T. parva* from cattle and not that *T. parva* of cattle was derived from a buffalo parasite. We think this because the buffalo and vector ticks existed in Central and South Africa together with cattle for many years before 1904 without recognizable theileriasis of cattle or buffalo. The disease, when introduced by cattle from East Africa, behaved like a disease that was new to the area.

CYTAUXZOON SPECIES IN ANTELOPES

The frequent occurrence of *Theileria*-like piroplasms in the red cells of wild antelopes has been mentioned earlier. Neitz (1964) has pointed out that these need not necessarily be *Theileria* species. On the contrary, he suggests that many, if not all, may eventually be identified as *Cytauxzoon* species. Unfortunately the genus *Cytauxzoon* is somewhat retiring and does not readily submit itself for study. Species have so far been described from three species of African antelopes and in each case it appeared that the parasite was the cause of death. It is, of course, probable that the parasite is much more common than these scanty records suggest for dead animals are quickly consumed by scavengers in Africa. The three species are *Cytauxzoon sylvicaprae* Neitz and Thomas, 1948, of the grey duiker (*Sylvicapra grimmi grimmi*); *C. strepsicerosi* Neitz, 1957, of the greater kudu (*Strepsiceros strepsiceros strepsiceros*) and *C. taurotragi* Martin and Brocklesby, 1960, of the eland (*Taurotragus oryx pattersonianus*).

In all cases schizonts and intra-erythrocytic piroplasms were described and in the case of *C. taurotragi* stages were found in sections of ticks collected from the dead eland. Brocklesby (1962) pointed out that there was no proof that these three stages belonged to the same parasite but believed that this was likely to be so: he was able to transmit a parasite to a bovine with *Rhipicephalus pulchellus* collected from the eland and demonstrated *Theileria*-like macroschizonts in smears of a lymph node but was unable to identify this parasite. So in the absence of tick-transmitted strains the only material available for study are tissues collected from fatal cases: we cannot even be sure that *Cytauxzoon* species are transmitted by ticks. The erythrocytic stages and the forms found in the salivary glands of the ticks were not really distinguishable from similar stages of *T. parva*, apart from a tendency for more definite cytomere formation in the early stages found in the ticks from the dead eland.

The most interesting forms were the schizonts and these, at least to an observer accustomed to *Theileria*, were truly bizarre. Schizonts of *C. sylvicaprae* and *C. strepsicerosi* occurred in histiocytes and formed large syncytial masses up to $660 \times 200 \mu$. These masses contained numerous cytomeres (or schizonts) which liberated hundreds of merozoites. The nucleus of the parasitized cell was stimulated to divide so that the resultant syncytium could contain 30 or more nuclei. In the duiker the most obvious and striking feature in the organs was the presence in the blood vessels of these large multi-nucleated syncytia that were normally attached to the inner wall of a blood vessel. Most of them were elongated and tailed off within the vessel lumen to a rounded or a pointed free end, floating in the blood stream. In arteries less than 1 mm in diameter anything from 1 to 50 of the bodies could be found in a single cross section.

Distribution of these schizonts was extremely variable. W. O. Neitz (1957 and 1960, personal communication) was able to find them in all the organs he examined from one dead duiker but only in sections of the adrenal gland in the other fatal case. In the kudu he could not find them in sections at all but only in smears prepared from spleen, liver and lymph nodes. Schizonts of the eland parasite were difficult to detect in smears, where they were indistinct, but were readily found in localized foci in sections of lymph nodes, lung and liver (Figs 1 and 2). In the eland schizonts were found developing in liver parenchyma cells: the host cells in other tissues could not be determined although it was felt that in the blood vessels of the lung they had probably arrived as emboli. The liver stages of *C. taurotragi* were often in the form of syncytia which measured up to $175 \times 68 \mu$ and bore some resemblance to the analogous stages of *Leucocytozoon* and *Hepatocystis*. This

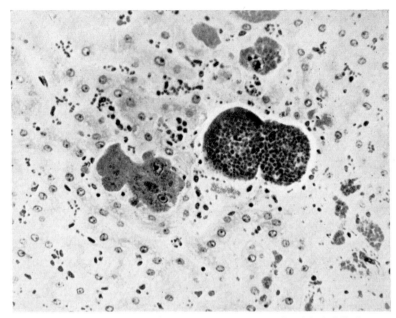

Fig. 1. Schizonts of *C. taurotragi* in liver. The parasite on the right is full of cytomeres and the one on the left is producing merozoites (× 500).

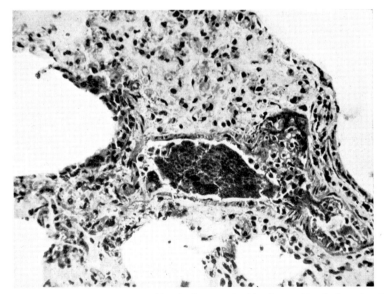

Fig. 2. Schizonts of *C. taurotragi* in the lumen of a blood vessel in the lung (× 500).

stimulated intriguing taxonomic speculations but Garnham (1966) concluded that it was doubtful if such similarities should be taken as evidence of taxonomic connection.

BABESIDAE OF CATS

These parasites are at present being studied by Dr. H. K. Dennig at the University of Munich and his results will be published in due course.

The literature is somewhat confused and will not be reviewed here. It seems to have been assumed that the Felidae are host to only one species of piroplasm; for instance, Levine (1961a) in his textbook "Protozoan Parasites of Domestic Animals and of Man" lists only one, *Babesia felis*, and gives *Babesiella felis* and *Nuttallia felis* var. *domestica* as synonyms.

"Elsa", the lioness featured in the series of books by Mrs. Joy Adamson, was released into the wild and died shortly afterwards. One of us (S.F.B.) examined blood smears and concluded that the cause of death was a piroplasm. This organism was a small form and was similar to one that was found (D.W.B.) in a lion shot by Dr. D. De Tray on the laboratory estate at the East African Veterinary Research Organization, Muguga, Kenya. Similar parasites were seen earlier in two lions from Uganda.

Dr. H. K. Dennig spent two years in Kenya and made preparations to isolate parasites from wild Felidae into splenectomized domestic kittens that had been born and reared under tick-free conditions. Unfortunately, this was not achieved until after he had returned to Germany. In October, 1964, Dr. Heather Campbell collected blood from a leopard that she was treating for a skin condition and this was inoculated into the domestic cats: this resulted in the isolation of a large *Babesia* species that was quite unlike the small piroplasm seen in Elsa (Brocklesby *et al.*, 1965). In March, 1965, after several failures with blood, an infected cat was sent to Dr. Dennig in Munich and he has established the parasite in experimental cats in his laboratory. Later in 1965 a second piroplasm was isolated from a leopard caught at Elmenteita in Kenya (D. W. Brocklesby and B. O. Vidler, unpublished). This parasite, which proved to be a small form like the one seen in the blood of Elsa, is also now under study in Munich. Dr. Dennig has also obtained a third parasite from a South American otter cat (*Herpailurus jaguarundi*) and he believes this to be distinct from the two piroplasms isolated from leopards; this has been the subject of a preliminary communication (Dennig, 1967).

So it is certain that the Felidae have at least two different piro-plasms. The question of nomenclature is difficult. The name *Nuttallia felis* (Davis, 1929) should be reserved for the small piroplasm that pro-duces Maltese cross forms. This was proposed by Carpano (1934) for the parasite originally named *Babesia felis* by Davis (1929). Jackson and Dunning (1937) evidently agreed with Carpano (1934) and went on to call their own parasite *Nuttallia felis* var. *domestica*. An entirely new name must be proposed for the large piroplasm of Felidae: this cannot be *Babesia felis* as this name was used by Davis (1929). We anticipate that the extensive studies being carried out by Dr. Dennig will solve this problem.

ECHINOZOON HOOGSTRAALI*

This is an extraordinary piroplasm found in 6 out of 12 rock hyrax (*Heterohyrax syriacus*) collected in Torit in the Equatoria Province of the Sudan by H. Hoogstraal and D. K. Lawless.

The parasite is remarkable because of the presence of filaments surrounding the infected erythrocytes. These filaments stained red with Geimsa and were usually $2.5\,\mu$ in length though some were twice as long and others little more than a dot. They appeared to arise from the surface of the red cell and were only found around cells that were infected with parasites over a certain size: there was a tendency for the first or smaller bodies to be associated with more filaments than the larger parasites. Older parasites seemed to have fewer filaments that tended to be more irregular in length, rather as though they had been battered in the rough and tumble of the circulation. The number of filaments was fairly uniform (16–30) and the majority of cells possessed close to the average number of 20–22. Red cells containing more than one parasite did not possess a proportionately greater number of filaments. The nature of the filaments was not established but Garnham (1951) thought that there was some suggestion that they might originate from the envelope of the erythrocyte. Certainly these filaments are very reminiscent of the filaments that one sees "growing" out of red cells in coverslip preparations of blood although these do not stain with Giemsa. The association between the parasitized cell and the filaments is certain but it might be still a peculiar attribute of the rock hyrax red cell and not of the parasite. If anyone has access to rock hyraxes, a study of the physiology of their red cells might prove interesting and might throw some light on this parasite. It would also

* Garnham (1951), a parasite of the Rock Hyrax.

be interesting to study the behaviour of parasites such as *Babesia microti* in rock hyraxes. If the filaments are not of parasitic origin the organism would presumably be re-classified in the genus *Babesia*.

*NUTTALLIA DANII**

The nomenclature of the *Nuttallia* species of *Meriones* is confused. It was discovered in 1930 by Adler and in 1952(a) he and Feldman-Muhsam were able to show that it was capable of multiplication in the erythrocytes of suckling mice. Tsur, Hadani and Pipano (1960) described the course of infection in the gerbil (*Meriones tristrami shawi*) and gave it the name of *Nuttallia danii*. Two years later Feldman-Muhsam (1962) described what is evidently the same parasite as *N. adleri*. A similar piroplasm of *Meriones erthyrourus*, the red-tailed gerbil, has been studied in Russia where it is referred to as *N. tadzhikistanica* (Krylov, 1963, 1964, 1965). Unfortunately we are not able to read these papers in the original. Cheissin (1966) has given a brief account of the life cycle of *N. tadzhikistanica* in *Hyalomma anatolicum*.

It would seem that the correct name of this organism is *Nuttallia danii* Tsur, Hadani and Pipano, 1960.

This parasite is of considerable interest for several reasons. It forms a link between the Babesidae and the Theileridae because *Theileria*-like schizonts occur in the internal organs, especially the liver and lungs. The type of cell parasitized by these schizonts has not been determined.

Small schizonts with 4 or 5 nuclei were 5 μ in diameter and schizonts with more than 100 nuclei were 35 × 17 μ. Krylov (1964) also described exo-erythrocytic stages: mono- and multinuclear forms lying free in the plasma in capillaries of the lungs, liver, myocardium, spleen and brain.

Feldman-Muhsam (1962) found stages in the salivary glands of infected nymphs that were reminiscent of the tick stages of *T. parva*.

A further reason for selecting this parasite for mention is that it can be transmitted experimentally by ticks. Adler and Feldman-Muhsam (1952b) were successful with *Rhipicephalus sanguineus* and Feldman-Muhsam (1958) with *R. secundus*; Tsur, Hadani and Pipano (1960) used *Hyalomma excavatum* and Krylov (1963) used *H. anatolicum*. He recommended the use of *N. tadzhikistanica* as a model for research in piroplasmoses. No transovarial transmission has been found.

---

* A parasite of the Gerbil.

Such a tick-transmitted piroplasm of small laboratory animals has been needed for many aspects of research and it is to be hoped that its use will be explored thoroughly by those in a position to do so. A. Hadani and E. Pipano (1968, personal communication) indicate that the organism might be very useful in drug screening procedures.

Discussion

We have chosen these piroplasms of wild mammals for several reasons. Firstly, because several of them stand astride the lines drawn by taxonomists and this emphasizes that these lines are purely artificial and should not be the subject of worship. Secondly, because we think that they have been neglected and that at least two of the parasite–host systems could well be used as useful models for research. Thirdly, to show how common the piroplasms are in wild mammals and to remind readers that some of them could be dangerous pathogens of man or his domestic animals. Fourthly, we hope that these remarks will make more people "piroplasm conscious" and include a box of microscope slides in their gear when they go collecting.

The finding of parasites in the odd blood smear, however, fascinating although it is for the moment of discovery, is always only a tantalizing glimpse; no more than the lift of a curtain. Always we yearn to know more and almost always we are frustrated: so we should try to establish some logical line of approach so that any discoveries that are made can be followed up by some systematic work. One line that can be explored without too much difficulty is to try to find out whether the piroplasms of wild creatures will infect domestic animals. For work on the wild animals themselves we would really need the resources of a large zoo with permission to commit wholesale splenectomy.

Since it is now plain that many of the piroplasms are far from being at all host specific, further extraordinary relationships will certainly be revealed in the future. We should first examine those ticks which are known to be capable of acting as vectors of piroplasms, see what host animals the various stages commonly infect and whether these animals harbour morphologically similar piroplasms.

Several times we have indicated how much work on piroplasms, especially the Theileridae, has been hampered by the lack of laboratory animal models. In the case of East Coast Fever, for instance, all research work has to be carried out using cattle as the experimental animals and these are expansive and cumbersome. In Italy a *Theileria* made a fleeting appearance in hares in the hunting grounds of Parma (Vaccari and Pieresca, 1958; Vaccari, Ballarini and Pieresca, 1961;

1962) but could not be isolated at that time: it was called *Theileria leporis* and the late Professor Vaccari made preparations hoping that it would reappear but apparently it has never done so. Anyone who has the opporunity to examine material from hares should certainly bear this parasite in mind. Probably the best place to make a systematic search would be Africa for there are several indications that the hares there are commonly infected with ticks of species that we know are capable of acting as vectors of other *Theileria* species. For instance Yeoman and Walker (1967) record that on 6 out of 14 Ukamba hares (*Lepus europaeus crawshayi*) in Tanzania nymphs of *Rhipicephalus appendiculatus*, the common vector of East Coast Fever, were found; two nymphs of *R. evertsi*, another vector of *T. parva*, were also found on one of the hares. Theiler (1962) also notes the presence of known *Theileria* vectors on hares in South Africa. These records, and others, suggest that a search for *T. leporis* in Africa could be fruitful.

Apparently innocent ticks should not be entirely neglected. An example of a tick with a virtually unblemished record is *R. pulchellus* which has so far only been shown to transmit Nairobi sheep disease. As mentioned earlier, stages of a *Theileria*-like piroplasm were found in the salivary glands of some *R. pulchellus* collected from a dead eland. This was a surprise and alerted us to the possibility that this tick might be a vector of some piroplasms after all. So we tried to transmit *T. parva* with it and found that it would indeed transmit this organism, albeit very inefficiently. Later we found more *Babesia*- and *Theileria*-like bodies in the salivary glands of ticks collected from a zebra that was carrying *N. equi* (Brocklesby, 1965). Further work will surely reveal that *R. pulchellus* is a natural vector of some piroplasms.

The acarologist has been pointing the way for the protozoologist for many years but, because they have often been buried in long lists in immense survey reports, their indications have been largely ignored.

Arthur (1966) recently presented his reasons for thinking that *B. microti* and *N. equi* might be the same parasite. They are similar morphologically and Maltese cross forms are commonly seen in both infections. At least 4 of the 7 known tick transmitters of *B. equi* feed on rodents or insectivores in their larval or nymphal stages and transovarial transmission has yet to be demonstrated in these 4 species. So at least these species of tick have the opportunity of acquiring infection from small rodents when they feed on them in their immature stages and of transmitting it when they feed as adults on horses. Furthermore, in Russia *N. equi* occurs north of 45°N but none of the transovarial tick transmitters extend north of this line. However, the rodent-feeding ticks do occur in these latitudes. Clearly, here is a pointer to some interesting work: a first step would be to see whether horses are susceptible to infection with *B. microti*. In Africa the situation may be even more interesting and complex for in addition to a heavy population of presumably infected small rodents there is a proven (Dennig, 1966) reservoir of infection in the wild zebra. Here it would be interesting to see if the *N. equi* in the zebra could be established in small laboratory rodents.

We have indicated that the concept of a strict host specificity in the piroplasms is gradually being eroded as more success attends attempts to transfer these parasites to new hosts. However, virtually all this work is performed using splenectomized animals as recipients. We therefore conclude by posing a question. Does the establishment of a hitherto exotic piroplasm in a splenectomized new host mean that this host has an inherent susceptibility? In other words, are there natural stresses that are equivalent to splenectomy?

ACKNOWLEDGMENTS

We are very grateful to Drs. A. Hadani and E. Pipano and to Professor W. O. Neitz who responded so willingly to our requests for information.

REFERENCES

Adler, S. (1930). The results of splenectomy in white mice as indicated by their reaction to *Bartonella muris*. *Trans. R. Soc. trop. Med. Hyg.* **24**, 75–82.
Adler, S. and Feldman-Muhsam, B. (1952a). Infection of Suckling Mice with a *Nuttallia* of *Meriones tristrami*. *Bull. Res. Coun., Israel* **2**, 201.
Adler, S. and Feldman-Muhsam, B. (1952b). Transmission of a *Nuttallia* of a Gerbil by *Rhipicephalus sanguineus*. *Nature, Lond.* **169**, 552.
Arthur, D. R. (1966). The Ecology of Ticks with Reference to the Transmission of Protozoa. In *"Biology of Parasites"*. 61–84. E. J. L. Soulsby (ed.). Academic Press, New York and London.
Barnett, S. F. and Brocklesby, D. W. (1966). The Susceptibility of the African Buffalo (*Syncerus caffer*) to Infection with *Theileria parva* (Theiler, 1904). *Br. vet. J.* **122**, 379–386.
Braff, E. and Condit, P. (1967). Morbidity and Mortality Weekly Report. (U.S. Dept. of Health, Education and Welfare, Bureau of Disease Prevention and Environmental Control, National Communicable Disease Center, Atlanta, Georgia, Jan. 7).
Brocklesby, D. W. (1962). *Cytauxzoon taurotragi* Martin and Brocklesby, 1960, A Piroplasm of the Eland (*Taurotragus oryx pattersonianus* Lydekker, 1906). *Res. vet. Sci.* **3**, 334–344.

Brocklesby, D. W. (1964). Parasites of the Family Theileridae of the African Buffalo occuring in East Africa. *Dr. Med. Vet. Thesis:* University of Zurich, 114 pp. (photo-litho).

Brocklesby, D. W. (1965). Evidence that *Rhipicephalus pulchellus* (Gerstacker, 1873) may be a Vector of some Piroplasms. *Bull. epizoot. Dis. Afr.* **13**, 37–44.

Brocklesby, D. W. and Barnett, S. F. (1966). The literature Concerning Theileridae of the African Buffalo (*Syncerus caffer*). *Br. vet. J.* **122**, 371–379.

Brocklesby, D. W., Dennig, H. K. and Vidler, B. O. (1965). A *Babesia* species of the Leopard (*Panthera pardus*) and its transmission to the domestic cat (*Felis catus*). *Int. Congr. Protozool.* No. 2, 177–178.

Brocklesby, D. W. and Vidler, B. O. (1966). Haematozoa found in wild Members of the Order Artiodactyla in East Africa. *Bull. epizoot. Dis. Afr.* **14**, 285–299.

Carpano, M. (1934). Sur les piroplasmoses des Carnassiers et sur un nouveau piroplasme des vélins (*Babesiella felis* chez le puma: *Felis concor*). *Bull. Minist. Agric. Egypt tech. scient. Serv.* No. 136, 20 pp.

Cheissin, E. M. (1966). The Systematical Position of Piroplasmids and some Peculiarities of their Development in the Vertebrate and Invertebrate Hosts. *Int. Congr. Parasit.* **1**, 259–260.

Davis, L. J. (1929). On a piroplasm of the Sudanese wild cat (*Felis ocreata*). *Trans. R. Soc. trop. Med. Hyg.* **22**, 523–534.

Dennig, H. K. (1966). The Isolation of *Babesia* species from Wild Animals. *Int. Congr. Parasit.* **1**, 262–263.

Dennig, H. K. (1967). Eine unbekannte Babesienart beim Jaguarundi (*Herpailurus yaguarundi*). *Kleintier-Prax.* **12**, 146–152.

Feldman-Muhsam, B. (1958). A note on the transmission of a *Nuttallia* of *Meriones shawi* by the tick *Rhipicephalus secundus*. *J. Protozool.* **5**, 225–226.

Feldman-Muhsam, B. (1962). On *Nuttallia adleri* of *Meriones tristrami*. *Int. Congr. Ent.* No. XI. **3**, 91–94.

Fitzpatrick, J. E. P., Cotton Kennedy, C., McGeown, M. G., Oreopoulos, D. G., Robertson, J. H. and Soyannwo, M. A. D. (1968). Human Case of Piroplasmosis (Babesiosis). *Nature, Lond.* **217**, 861–862.

Garnham, P. C. C. (1951). A new Piroplasm from the Rock Hyrax. *J. Parasit.* **37**, 528–532.

Garnham, P. C. C. (1966). Malaria Parasites and other Haemosporidia. Blackwell Scientific Publications, Oxford, p. 32.

Garnham, P. C. C. and Bray, R. S. (1959). The Susceptibility of higher primates to piroplasms. *J. Protozool.* **6**, 352–355.

Honigberg, B. M., Balamuth, W., Bovee, E. C., Corliss, J. D., Gojdics, M., Hall, R. P., Kudo, R. R., Levine, N. D., Loeblich, A. R., Weiser, J. and Wenrich, D. H. (1964). A Revised Classification of the Phylum Protozoa. *J. Protozool.* **11**, 7–20.

Jackson, C. and Dunning, F. J. (1937). Biliary fever (Nuttalliosis) of the cat: A case in the Stellenbosch District. *Jl S. Afr. vet. med. Ass.* **8**, 83–88.

Krylov, V. M. (1963). (The red-tailed Gerbil, *Meriones erythrourus*, as an experimental model for the study of piroplasmosis.) *Mater. Kong. Probl. Prot.* 61–62.

Krylov, V. M. (1964). (Development of some piroplasms within the vertebrate hosts.) *Acta prot., Warsz.* **2**, 307–329.

Krylov, V. M. (1965). (An experimental model for the study of piroplasmosis: *Nuttallia tadzhikistanica* infection in the red-tailed gerbil, *Meriones erythrourus*.) *Veterinariya* **42**, 52.

Levine, N. D. (1961a). Protozoan Parasites of Domestic Animals and of Man. Burgess Publishing Company, Minneapolis, Minnesota, U.S.A.

Levine, N. D. (1961b). Problems in the Systematics of the "Sporozoa". *J. Protozool.* **8**, 442–451.

Martin, H. and Brocklesby, D. W. (1960). A New Parasite of the Eland. *Vet. Rec.* **72**, 331.

Neitz, W. O. (1955). Corridor Disease: A fatal form of bovine theileriosis encountered in Zululand. *Bull. epizoot. Dis. Afr.* **3**, 121–123.

Neitz, W. O. (1957). Theileriosis, Gonderioses and Cytauxzoonoses: A Review. *Onderstepoort J. vet. Res.* **27**, 275–430.

Neitz, W. O. (1964). Cytauxzoonoses in African Antelopes. *Bull. Off. int. Epiz.* **62**, 1009–1015.

Neitz, W. O. and Thomas, A. D. (1948). *Cytauxzoon sylvicaprae* gen. nov., spec. nov., a Protozoon Responsible for a Hitherto Undescribed Disease in the Duiker (*Sylvicapra grimmia* (Linne)). *Onderstepoort J. vet. Sci. Anim. Ind.* **23**, 63–76.

Riek, R. J. (1964). The development of *Babesia bigemina* (Smith and Kilborne, 1893) in the tick *Bosphilus microplus* (Canestrini). *Aust. J. agric. Res.* **15**, 802–821.

Shortt, H. E. and Blackie, E. J. (1965). An Account of the Genus *Babesia* as Found in Certain Small Mammals in Britain. *J. trop. Med. Hyg.* **68**, 37–42.

Skrabalo, Z. and Deanovic, Z. (1957). (Piroplasmosis in Man). *Doc. Med. Geograph. Trop.* **9**, 11–16.

Theiler, G. (1962). African Ticks: The Complexity of their Host-Encounter Relationship: Disease and Control. *2nd. Meeting FAO/OIE Expert Panel on Tick-Borne Diseases of Livestock, Cairo (FAO Meeting Report AN 1962/10)*, 45–52.

Tsur, I., Hadani, A. and Pipano, E. (1960). *Nuttallia danii* n.sp.—A Haemoprotozoon from the Gerbil (*Meriones tristrami shawi*). *Refuah vet.* **17**, 244–236.

Vaccari, I., Ballarini, G. and Pieresca, G. (1961). Theileriasi delle lepri—"Corpi di Koch" in *Lepus europaeus* Pall. *Nuova Vet.* **37**, 1–7.

Vaccari, I., Ballarini, G. and Pieresca, G. (1962). Theileriasi delle lepri—Equivalenze e rapporti limitari ematici in individui di *Lepus europaeus* Pall. ammalati e clinicamente sani. *Ric. Zool. appl. Caccia Suppl.* **4**, 87–93.

Vaccari, I. and Pieresca, G. (1958). Theileriasi delle lepri—*Theileria leporis* in *Lepus europaeus* Pall. *Nuova Vet.* **34**, 1–10.

Yeoman, G. H. and Walker, J. B. (1967). The Ixodid Ticks of Tanzania. *Commonwealth Inst. Ent., London*, 214 pp.

DISCUSSION

YEOMAN: Neither the hypothesis that cattle are the natural biological host of *Theileria parva* nor that buffalo are seems entirely satisfactory. Although cattle are good hosts of *Rhipicephalus appendiculatus*, that they are not ideal hosts of immature stages is suggested by the finding that they seem to carry lower populations of immature stages than adults. In general, small mammals are better hosts of the immature stages of Rhipicephalids. Was there a case for suggesting that there might exist a so far unrecognized small animal reservoir of *T. parva* which is maintained by larval-nymphal transmission?

BARNETT: The efficiency of the transmission of *T. parva* had always been a puzzling feature of the disease because of the apparent absence of carrier cattle as a source of infection and, as Mr. Yeoman has found, the relatively small numbers of immature stages of *R. appendiculatus* found on cattle. The carrier animal may be more frequent than was thought and this, together with the high efficiency of the nymphal to adult infection in the tick, might be sufficient to provide the high incidence of infection in the field. Nevertheless a search for possible small mammal reservoirs of *T. parva* should rank high in the priorities of research on the epidemiology of *T. parva* infection.

Symp. zool. Soc. Lond. (1968) No. 24, 177–195.

BLOOD PROTOZOA OF FREE-LIVING BIRDS

CARLTON M. HERMAN

Patuxent Wildlife Research Center,
Laurel, Maryland, U.S.A.

SYNOPSIS

Blood protozoa were first reported from wild birds in 1884. Since then numerous surveys throughout the world have demonstrated their presence in a wide variety of hosts and localities with continuing designations of new species. Taxonomic determinations include parasites in the genera *Plasmodium*, *Haemoproteus*, *Leucocytozoon*, *Babesia*, *Lankesterella* and *Trypanosoma*.

Transmission of *Plasmodium* by mosquitoes was demonstrated with a bird parasite before these insects were proven as vectors of human malaria. All the genera under consideration require an insect vector to complete their life-cycles and susceptible vectors have been demonstrated. Most experimental work on the blood protozoa of birds has been carried on with captive birds. An extensive volume of research has been conducted on *Plasmodium* because of its close similarity to malaria in man. Field studies that would provide information on the epizootiology of occurrence of these parasites in wild populations have been very limited, mainly confined to single blood film surveys. Such data are inadequate to provide an understanding of true prevalence or incidence or of factual knowledge of their impact on the wild population. Mechanisms for procuring such information are available in some cases and can be developed to fit other situations. Isodiagnosis, inoculation of blood from wild birds into susceptible captive hosts, has revealed a prevalence of over 60% for *Plasmodium* in situations where microscope examination of single peripheral blood preparations yielded less than 1%. Culture of bone marrow collected by biopsy demonstrates high prevalence of trypanosomes even when none are evident from microscopic examination of blood. Often preparations of tissues collected at necropsy reveal *Leucocytozoon* and *Lankesterella* when examination of peripheral blood gave no indication of infection. Methods developed by bird ringers provide techniques for obtaining repeat examinations of free-living birds that can yield further pertinent data. The use that has been made of these techniques, beyond the mere procurement of a single blood film, is reviewed from the standpoint of providing knowledge of the epizootiology of the blood protozoa of birds and their further potential is projected.

INTRODUCTION

The protozoa that can be observed, at some stage in their development, in preparations of the blood of birds include the genera *Plasmodium*, *Haemoproteus*, *Leucocytozoon*, *Babesia*, *Lankesterella* and *Trypanosoma*. All but the trypanosomes belong to the Sporozoa and occur in blood cells. Since the first report of a protozoan from the blood of birds by Danilewsky (1884), there has been a constantly growing list of species and hosts. The earlier investigations and reports were excellently reviewed and summarized by Wenyon (1926). Since Wenyon's classical contribution, the taxonomy of most of these forms has been reviewed and expanded by many investigators, including Manwell (1935, 1938); Giovannola (1939); Hewitt (1940); Corradetti, Garnham and Laird (1963); Lainson (1959) and others.

At least 54 species of *Plasmodium* from birds have been named. It is not the purpose of this presentation to discuss the validity of any of these species. Undoubtedly many of them are synonyms. Attempts also have been made to categorize some of these forms into subgroups or subgenera (Corradetti *et al.*, 1963; Garnham, 1966). Procedures required for a definitive diagnosis usually go beyond the mere microscopic examination of a single blood film, requiring information on host range, periodicity, vector range and other factors that can be clarified only by extensive laboratory observations. Ross (1898) demonstrated transmission of a bird *Plasmodium* before mosquitoes were known to transmit malaria in man. Transmission is usually by culicine mosquitoes although, on occasion, anophelines have proven susceptible.

The number of species of *Haemoproteus* and *Leucocytozoon* named is also very extensive. The validity of many of these is in even greater confusion than is the genus *Plasmodium* for many have been reported with very limited morphological description, the chief criterion being the fact that the parasite had not been previously reported from a particular host species. There are no data to warrant the conclusion that these parasites have such a tight host specificity as to justify this approach. Bennett, Garnham and Fallis (1965) attempted to subdivide these genera in some sort of a logical arrangement but only a limited number of the species have been considered. The data available on most species named are too scanty to permit any categorization. Some species of *Leucocytozoon* are transmitted by blood-sucking blackflies (Simuliidae). At least one is known to be transmitted by *Culicoides*. Hippoboscid flies have been demonstrated as vectors of some species of *Haemoproteus* and *Culicoides* of others.

Babesia has been reported infrequently from avian hosts. Its taxonomic status was reviewed recently by Laird and Lari (1957). Very few species are recognized. Transmission is by argasid ticks.

Toxoplasma has, on many occasions, been reported from birds on the basis of morphological identification of stages in blood cells or tissues. Although true *Toxoplasma* has been demonstrated in avian hosts, most of the forms reported in earlier literature apparently belong to a different group. Recognizing this confusion, Garnham

(1950) named these forms *Atoxoplasma* and, ultimately, Lainson (1959) proposed that they belong to the genus *Lankesterella*. Mites serve as the vectors.

The trypanosomes of birds, as far as is known, are confined to body fluids and are not associated with blood cells or tissue cells. Their extreme pleomorphism has led many workers to consider the bird-inhabiting forms monospecific *Trypanosoma avium*. However, other investigators have followed the procedure of naming new species on the basis of finding them in previously unreported hosts, assuming again a tight host specificity. Bennett (1961) has demonstrated that such host specificity does not exist and that wide variations in size are evident in experimentally established infections. Thus morphological descriptions based primarily on size or shape are of questionable validity.

Blood parasites have been reported for a wide variety of wild birds from many parts of the world. Many host lists have been published, among which might be mentioned: Levine and Kantor (1959), Coatney (1936, 1937), Coatney and Roudabush (1936, 1949), Herman (1944), Berson (1964) and Lucena (1939).

Most of these records are based on examinations of single blood films. Numerous surveys have been conducted in many parts of the world but mostly in temperate areas, undoubtedly because of the presence of a greater abundance of investigators. These surveys were undertaken chiefly to determine the distribution of parasites in different hosts, to add to our knowledge of geographic and host associations of the parasites and in the hope of finding new forms. Many investigators have attempted to present infection ratios and to interpret incidence or prevalence on the basis of their survey data. In most cases little significance can be placed on such interpretations.

Figure 1 illustrates a graphic representation of the parasitemia that occurs in most of the sporozoan infections. The prepatent period is the time interval from exposure to evidence of parasites in the blood; the patent period is the interval during which parasites are apparent in the blood. If, at times, parasitemia is recognized during latency, it is usually only during a relapse. The duration of the patent period varies considerably, particularly in naturally occurring infections. Such data are obtained from repeat samples taken from retrapped birds or from birds in captivity.

PLASMODIUM

Because *Plasmodium* can be readily transmitted by subinoculation of blood into a susceptible host, much more is known of the biology of this genus in avian hosts than of the other genera of parasites under discussion. During the two decades 1920–1940, the canary was the chief experimental host and the species of *Plasmodium* studied originated from passeriform birds. The isolation of species from larger avian

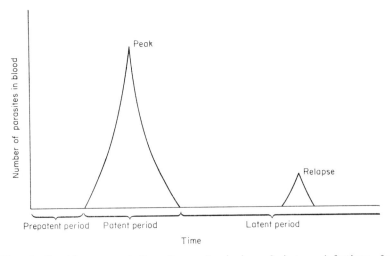

Fɪɢ. 1. Graphic representation of parasitemia in malaria-type infections; from Herman (1968).

hosts (*P. gallinaceum, P. lophurae, P. fallax, P. durae, P. juxtanucleare*) initiated extensive studies in domestic chickens, ducks and turkeys. Duration of patency may last a single day but usually parasitemia is evident from one to several weeks. *Plasmodium* infections follow the pattern of the graphic presentation of parasitemia much more closely than do those caused by the other generic groups. Thus it can be seen that unless survey samples are taken during the limited period of initial patency, the chance of recognizing infected birds is minimal. Highest incidence occurs in parallel with the activity of vectors.

It appears that once infected a bird remains infected for life. A number of investigators have demonstrated the ability to transmit avian *Plasmodium* spp. after years of latency by subinoculation of blood into a susceptible host. Several investigators have utilized this phenomenon in surveys of prevalence of *Plasmodium* when the routine blood examination revealed no organisms. I employed this technique in studies on the epizootiology of *Plasmodium* in red-winged blackbirds (*Agelaius p. phoeniceus*) to good advantage in Massachusetts over 30 years ago (Herman, 1938). By using canaries as recipients at each recapture of wild young birds of the year, I was able to demonstrate that 60% of the young were already infected before their southward migration and that they did not become infected until some time after leaving their nests. Similar studies conducted subsequently with house finches and sparrows in California indicated that much of the infection in these species was obtained while they were still in the nest (Herman, Reeves, McClure, French and Hammon, 1954). Most recently we employed this isodiagnosis technique in some studies on Canada geese (*Branta canadensis*) in northern Michigan (Herman, Knisley and Snyder, 1966). One ml of blood was subinoculated from wild birds caught in July into 5-day-old domestic geese, which were subsequently sampled 3 times a week for the ensuing 2 months. Examination of blood films from the donor birds revealed no parasites but infection with a *Plasmodium* was diagnosed in approximately 60% of the recipients. This procedure was repeated for 4 successive years (Herman, 1968) with almost exactly the same percentage of prevalence each year. Survey data obtained by isodiagnosis are therefore likely to be much closer to the actual situation of prevalence of *Plasmodium* than data obtained by single blood film examinations.

However, other factors can confuse the picture. Most of our goose isolates have been diagnosed as *P. circumflexum*. Infrequently other species such as *P. relictum* have been observed. In the original recipient domestic goose, the parasitemia was at a level of one parasite for each 2000 to 5000 RBC. Subsequent subpassage into domestic geese demonstrated parasites in only 30% of the recipients. Sixteen months after the original isolation, we sub-passaged blood from the first infected domestic goose into a 3-day-old Pekin duck which subsequently developed a *P. circumflexum*-like infection. Although we have been able to carry this infection in ducklings, we have been unable to re-establish it in the geese.

We have experienced another interesting episode in our studies with *P. circumflexum*. We obtained canary blood, infected with *P. circumflexum* from a mynah bird in Ceylon, through the courtesy of Dr. Dissanaike when he was at the London School of Tropical Medicine a few years ago. This was readily passaged to domestic geese but several attempts to put it back into canaries have been unsuccessful.

Choice of recipient species of host is an important consideration in conducting or interpreting isodiagnosis data and results such as those just outlined simply add to the confusion. As with interpretation of results from surveys involving only single blood film examination, positive results provide usable data, negative results may be meaningless.

The diagnosis of a *Plasmodium* parasite from a wild bird tells us nothing of its potential as a mortality factor. Every investigator who has conducted transmission experiments in his laboratory by blood inoculation is aware that heavy infections may kill his animals. Visible signs in a bird can usually be described as a simple malaise or weakness, manifested largely by its drooping wings. Blood is likely to become thin and watery. Necropsy reveals a greatly enlarged, dark spleen. Histologic findings vary somewhat with species. Exoerythrocytic stages, as well as parasitized red cells, are likely to cause tissue destruction or blockage of capillaries. The type of pathology that may occur with each species has been reviewed very recently by Garnham (1966) in his book on malaria parasites and there is no need to repeat the details here.

What occurs in *Plasmodium* infections among wild birds under "natural" conditions is extremely difficult to determine with the limited techniques we have thus far developed. As pointed out by Corradetti, Neri and Scanga (1960), a passeriform bird brought into the laboratory will occasionally exhibit a severe relapse and succumb, presumably, from the infection. We have observed this on several occasions but we also have had many die presumably from the conditions of captivity such as injury or starvation, without evidence of parasites. Even if one performs a detailed necropsy, interpretation may be difficult.

In our initial studies on the Canada geese cited earlier, we found that about 10% of our domestic geese inoculated from wild geese infected with *P. circumflexum* died with severe parasitemia. In prior studies, we inoculated Canada goslings and the 6 that became infected all succumbed with a severe parasitemia. Several hundred mallard and Pekin ducklings infected during the course of our studies all survived. The extent of parasitemia appears to have a direct bearing on survival. When we inoculate very young domestic geese with large numbers of parasites, the recipients invariably succumb to a severe parasitemia. In such infections there may be many more parasites than there are red blood cells. Although we have examined blood films from several thousand Canada geese at all seasons and from a number of areas, we have only recognized a *Plasmodium* on half a dozen occasions and the number of parasites present was less than one per 2000 RBC. We have never examined any wild geese at the peak of patency and one might conclude they never get the severe parasitemia we generally observe in our blood-transfused infections in experimental birds. The high percentage of infected birds recognized by isodiagnosis represents only those donor birds that survived the initial onset. Whether or not any of the wild population of geese succumbed to *Plasmodium* infection is yet to be determined. To my knowledge no investigator has uncovered positive evidence of epizootic losses in any species of wild bird in nature. However, the history of *P. durae* in turkeys in Kenya (Purchase, 1942), of penguins in zoo collections (Fiennes, 1967) and the vast amount of experimental evidence from laboratory experience with a number of captive host species, together with the high prevalence demonstrated in some surveys of wild birds, certainly suggests a potential for severe epizootics.

Experimental studies conducted in Hawaii a few years ago tend to emphasize this potential. Mosquitoes were unknown from the Hawaiian Islands until the advent of white explorers. Shortly thereafter many of the indigenous lowland species of birds became extinct. Warner (1968) obtained certain species of finches from higher altitudes and from another Pacific island where mosquitoes were not known to occur and exposed them in the lower altitudes where they quickly succumbed with severe parasitemias of *Plasmodium*. This experience, as well as experiences with the exotic penguins in the temperate zone, seem to indicate that indigenous species may have an immunity to fatal disease even though we can demonstrate a high rate of infection.

It seems to me that studies on zoological collections, as well as more intensive study of wild populations, provide an opportunity to obtain answers to some of the questions implied by my remarks. *Plasmodium* is often a problem among penguins in north temperate areas. Fiennes (1967) has suggested that these birds become infected while on exhibit and we have come to the same conclusion in observations at the zoos in Baltimore and Washington (Herman, Kocan, Snyder and Knisley, 1968). By isodiagnosis we have uncovered several infections of *Plasmodium* in both native and exotic ducks in the collection of the National Zoo in Washington, D.C. We are attempting to determine the source of these parasites.

HAEMOPROTEUS

Huff (1932) pointed out that parasites belonging to the genus *Haemoproteus* are the most common malaria parasites of birds and yet they are probably the least known. He explained this state of affairs by the fact that no simple and efficient method exists for transferring the infection from bird to bird and hence laboratory strains cannot easily be maintained. The asexual forms in this genus parasitize fixed tissue cells instead of circulating blood cells as in the genus *Plasmodium*. Most studies have been conducted with species transmitted by hippoboscid flies in such hosts as Columbiforms and California quail.

Haemoproteus is a common parasite of ducks (Herman, 1954) and grouse, (Fallis and Bennett, 1960) but hippoboscid flies do not occur on Anatidae. In numerous discussions in the past I pointed this out but it remained for Fallis and Wood (1957) to demonstrate that certain Ceratopogonid flies are the vectors of this genus in ducks.

While the pathologic findings are quite similar to those in *Plasmodium* infections, particularly the enlarged dark spleen, reports of bird fatalities from *Haemoproteus* are rare. In studies of this parasite in California quail, over a period of years, the only quail which I could claim died of the infection was a captive young bird experimentally infected by a hippoboscid (*Stilbometopa*). This bird died with a severe parasitemia 10 days after initial appearance of parasitemia and 30 days after initial exposure.

Most extensive studies have been on Columbiform birds. *Haemoproteus* is a very common parasite of pigeons (*Columba livia*) throughout most of the world (Knisley and Herman, 1967). It also is common in mourning doves in many parts of the United States (Hanson, Levine, Kossack, Kantor and Stannard, 1957) and has been reported from the standpoint of taxonomy, host distribution and frequency of occurrence and vectors. Other factors have been studied little. Data on its possible impact on the host are lacking.

The course of parasitemia in *Haemoproteus* does not always adhere as closely to the graphic pattern as it does in *Plasmodium*. Often throughout the latent period, small numbers of parasites can be found in the peripheral blood. Frequently there is a high prevalence in May and early June. Bennett and Fallis (1960) have suggested that this increase in parasitemia could be the result of a relapse or acquisition of a new infection. In studies on naturally infected California quail maintained in captivity for periods up to nearly 5 years and examined 3 times weekly, parasites could be demonstrated continuously, ranging from 1 or 2 parasites per 10 000 RBC to as many as over 500 per 10 000 RBC (Herman and Bischoff, 1949). These birds were maintained free of hippoboscids which are the vectors of *H. lophortyx* in this host. Highest levels of parasitemia were observed in February and March.

Haemoproteus in pigeons can be found during all seasons. Bennett and Fallis (1960), in surveys of blood parasites in birds in Canada, found the highest prevalence in June and July, presumably corresponding with the time of greatest abundance of suitable vectors. In studies conducted in Maryland, I have found native adult wood ducks infected in April (Herman, 1954) but adults sampled in June and July revealed no parasites. However, in mallard ducks exposed to natural infection for periods of one week and subsequently confined in the laboratory, infections became evident in June and July. Parasitemia lasted only a few days and no parasites could be observed in subsequent examinations.

While the limited knowledge available on *Haemoproteus* in many respects parallels some of the findings in more extensive studies on *Plasmodium* and *Leucocytozoon*, much more detailed knowledge on the epizootiology of this group is needed before we can hope to recognize its potential as a pathogen.

LEUCOCYTOZOON

More is known of the impact of *Leucocytozoon* on some species of birds than of that of any of the other parasites under discussion. Borg (1953) reviewed the status of this parasite as a mortality factor in a number of previously reported cases. While admitting that *L. smithi* of domestic turkeys and *L. simondi* of ducks had produced acceptable evidence as causes of death, he was not convinced that *Leucocytozoon* was involved as a killer of grouse in Sweden except possibly under unusual circumstances. Knuth and Magdeburg (1922, 1924) attributed death to *Leucocytozoon* in young domestic geese one year but noted that infected birds in a subsequent year survived. An epizootic in weaver birds in Africa was attributed to a species of *Leucocytozoon* (Garnham, 1950). *Leucocytozoon* has been reported from a wide variety of avian hosts and, as with *Haemoproteus*, many species have been named primarily because they occurred in new hosts. Waterfowl known to be susceptible are cited by Herman (1963) and Fallis and Bennett (1966).

The most extensively studied species of *Leucocytozoon* in wild birds is *L. simondi*. Although first reported from a teal duck from south eastern Asia (Mathis and Leger, 1910), extensive studies on this parasite have been conducted mainly in an area including Michigan and Wisconsin in the United States and Ontario in Canada (Herman, 1968). It was first reported from this area by O'Roke (1934) who claimed it caused drastic mortality in ducklings, sometimes up to 100%.

In temperate regions, *Leucocytozoon* prevalence is highest in May, June and July (Borg, 1953; Bennett and Fallis, 1960). In waterfowl, a relapse occurs approximately at the time of egg-laying (Chernin, 1952). This proves very appropriate in supplying an adequate source of parasites for vector blackflies to carry the infection to the newly hatched susceptible young. Chernin (1952) exposed domestic ducks in Michigan during the summer months and observed two peaks of high transmission

(Table I). These data were obtained by exposing birds at weekly intervals. In a similar manner, Fallis and Bennett (1966) exposed domestic ducks at Algonquin Park in Ontario at different times from May to September over a period of 10 years. Transmission activity did not show as marked a periodicity as reported by Chernin and percentage of deaths varied from year to year although most birds developed a parasitemia when exposed in May, June and July.

At the Seney National Wildlife Refuge in the upper peninsula of Michigan, we have been conducting field studies on the occurrence of *Leucocytozoon* in Canada geese. This goose population was established in the late thirties and has built up to about 800 birds which migrate south for the winter in November-December and return in the spring in March. Young hatch in mid-May. About every fourth or fifth year die-offs occur in the goslings and in one year over 89% of the hatch of young was lost, chiefly from *Leucocytozoon*. This study has been in progress nearly 20 years and we have found the parasites in young birds every year. Severe infections are noted in the youngest birds. One hypothesis we have propounded is that birds infected in nature when less than one week old will usually succumb, while birds bitten by infected flies when more than two weeks old will usually survive the infection. Older birds with a larger initial inoculum of sporozoites could also succumb. In preliminary experiments, R. M. Kocan (unpublished) gave the same dosage of sporozoites obtained from *Simulium rugglesi* to ducks of varying ages. The dosage he used killed ducklings 1, 2 and 3 weeks old. Birds 4 and 5 weeks old developed a severe parasitemia but survived. When the same dosage was inoculated into birds 2, 3, 4 and 5 weeks old after recovery from an initial infection, no deaths occurred. We do not know how this dosage of sporozoites compares with the dosage that occurs in nature but we assume it is much greater. At any rate, we believe that age resistance and acquired immunity play a combined rôle in birds which recover from *L. simondi* infections. If adverse weather conditions exist at the time the young hatch, and fly biting is inhibited, fatalities in the goslings will be limited.

Studies on captive survivors, both ducks and geese, show that, after the initial crisis, the parasites virtually disappear from the blood. The patent period lasts but a few weeks. Extensive search of blood films reveals a few parasites throughout the winter. Parasites reappear in the blood in late April or early May in readily observable numbers but are never as abundant as in the initial phase of parasitemia. In our examinations of adult birds during eclipse plumage in early July, when we have bled them for *Plasmodium* studies, we have only rarely observed *Leucocytozoon* parasites in the blood films. Thus, if an

TABLE I

Morbidity, 10-day patency and case fatalities among groups of ducks exposed to natural infection with Leucocytozoon simondi *(summer 1950). (All ducks 6 weeks of age when exposed. Each group (I–VI) exposed for 8 days)*

Group designation	No. ducks exposed	Dates of exposure	Ducks infected		Ducks patent 10 days P.E.*		Case fatalities	
			No.	%	No.	%	No.	%
"Permanent"	11	June 16–Aug. 24	11	100	0	0	0	0
I	11	June 29–July 7	9	82	1	11	0	0
II	12	July 11–July 19	12	100	12	100	10	83
III	15	July 19–July 27	14	93	9	64	2	14
IV	10	July 27–Aug. 4	9	90	8	89	5	56
V	10	Aug. 8–Aug. 16	3	30	1	33	0	0
VI	10	Aug. 16–Aug. 24	0	0	0	0	0	0
Totals or averages**	68	June 29–Aug. 24	47	69	31	66	17	36

* P.E. = Post exposure.
** Totals do not include the "permanent" group which is treated in this analysis as an indigenous population.
From Chernin (1952).

investigator were to conduct a survey of this population in the winter, or even in July, there would be no evidence of a severe *Leucocytozoon* problem.

A number of investigators, after obtaining lower parasitemia with little mortality in experimental birds exposed for longer periods, have concluded that the birds are repeatedly infected and thereby develop an immunity. Although Chernin's (1952) table shows active periods of transmission in late June and July and Fallis and Bennett (1966) have demonstrated active transmission from late May through July, the events at Seney seem to indicate that the important transmission period is from mid-May to early June. However, in our efforts to obtain infected flies for a supply of sporozoites for laboratory experiments, we have noted a high prevalence of infected *S. rugglesi* well into July.

Although goslings that succumb to the infection may have a severe parasitemia by the time of death 4 to 11 days after initial exposure, we have noted an occasional bird with no parasitemia but with extensive involvement with tissue stages of the parasite. Kocan and Clark (1966) have reported on the anemia associated with severe infections of *L. simondi*. They suggest that death of young wild waterfowl may result from the summation of several stressing events in addition to the anemia, including direct blood loss to feeding by black flies and possibly from an auto-immunity and population interactions. In addition, we have sometimes found the geese at Seney to be heavily infected with other parasites, particularly *Amidostomum* and *Tetrameres*, and each year several goslings succumb to aspergillosis. These other parasites do not appear to be unusually abundant in the years of extensive losses, however.

Early this year, I pointed out some factors involved in the distribution of *L. simondi* in North America (Herman, 1968). Figure 2 illustrates the apparent southward limit of active transmission, roughly at about 43 degrees latitude. This geographic limitation appears to me to be an interesting facet of our knowledge. Further analysis should show this to be related to availability of suitable vectors or infected carriers or both.

BABESIA

Laird and Lari (1957) found a parasite in a single crow in India. In an effort to diagnose their findings, they reviewed the literature on similar forms reported from birds and decided these should all, along with their form, be classified as *Babesia*. The most common of such parasites reported from birds is *Aegyptianella pullorum*, originally described from fowl in Egypt and since reported from several other species. Although Laird and Lari presented much discussion of the

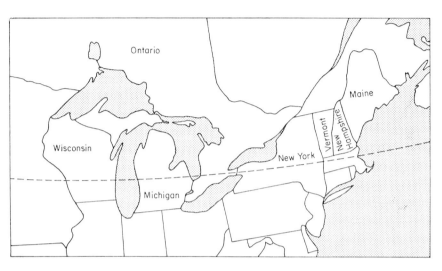

Fig. 2. Limit of *Leucocytozoon simondi*. Transmission appears to occur primarily north of 43° latitude.

taxonomic position of this parasite, which they apparently considered to be a *Babesia*, they did not specifically reduce *Aegyptianella* to synonomy.

Although *Aegyptianella* has been most frequently reported from chickens, it also occurs in other birds. Ahmed and Elsisi (1966) reported clinical cases in an extensive study of chickens in Egypt; however, the percentage of chickens infected was very low. Their figures on prevalence are confusing. They reported 643 cases of both *A. pullorum* and spirochetes in 5982 birds examined between January 1963 and June 1965, with a frequency ratio of 2 *A. pullorum* cases to 98 with spirochetes.

Since the publications of Laird and Lari, I have observed a *Babesia* in a crow (*Corvus brachyrhynchos*) from northern Michigan but the level of parasitemia was too low to determine if it was the same as that reported in the Indian crow. Over 20 years ago, in survey blood films of California quail, I frequently observed a *Babesia* so like that described by Laird and Lari that they could easily have prepared their illustrations from my material and obtained the same pictures. The infected quail showed no clinical signs and no ticks were found.

LANKESTERELLA

Certain parasites of birds were not recognized as *Lankesterella* until very recently (Lainson, 1959). This parasite is very common in house sparrows (*Passer domesticus*). Parasites I diagnosed from this host and other species of North American birds as *Toxoplasma* undoubtedly belonged to this genus (Herman, 1937). Many early records of *Hepatozoon*, *Haemogregarina* and *Toxoplasma* from birds are probably in the same status. *Lankesterella* apparently occurs in a wide range of passeriform birds in many parts of the world.

Box (1967) has suggested that presence of the coccidian *Isospora* in house sparrows produces increased parasitemia and death from *Lankesterella*.

TRYPANOSOMA

Shortly after the turn of the present century, Novy and MacNeal (1904) found that trypanosomes could be demonstrated by culture of blood from birds. Trypanosomes occur extremely rarely in stained blood films. Diamond and Herman (1954) demonstrated a prevalence rate of 40% in cultures from bone marrow of Canada geese, even although stained peripheral blood films of the same birds revealed no parasites. Stabler, Holt and Kitzmiller (1966) further emphasized that bone marrow was the preferred site for cultural isolation of avian trypanosomes. Bennett (1961) developed a technique of concentrating blood by centrifugation in a capillary pipette which yields a much higher prevalence of positive cases by examination of a fresh unstained preparation. This technique or culture of bone marrow demonstrates a much greater prevalence of trypanosomes in birds than stained preparations of peripheral or heart blood would suggest.

Studies by Bennett (1961) cover the transmission and morphological relationships of trypanosomes of birds from Ontario. Similar studies are needed elsewhere to clarify the specific classification of avian trypanosomes and, indeed, to fully demonstrate whether more than one species occurs in birds and what criteria are valid for their differentiation. Although several trypanosomes that occur in man and other mammals are highly pathogenic, no such relationship has been demonstrated for any trypanosomes infecting birds.

CONCLUSION

In conclusion, I wish to emphasize that most of the studies which have been reported on the blood protozoa of birds have been aimed toward an understanding of the biology of the parasites. Very little has been presented on the effect of blood protozoa on the host or host populations. A broad and challenging field awaits the ingenuity of future investigators. Maintenance of zoological park collections, natural park collections and the conservation of wildlife species are sorely in need of such data and present trends predict they shall be much more important in the future.

REFERENCES

Ahmed, A. A. S. and Elsisi, M. A. (1966). Observations on Aegyptianellosis and Spirochaetosis of Poultry in Egypt. *Egypt Med. J., Cairo University,* XI, 139–146.

Bennett, G. F. (1961). On the specificity and transmission of some avian trypanosomes. *Can. J. Zool.* **39,** 17–33.

Bennett, G. F. and Fallis, A. M. (1960). Blood parasites of birds in Algonquin Park, Canada and a discussion of their transmission. *Can. J. Zool.* **38,** 261–273.

Bennett, G. F., Garnham, P. C. C. and Fallis, A. M. (1965). On the status of the genera *Leucocytozoon* Ziemann 1898 and *Haemoproteus* Kruse, 1890 (Haemosporidiidae: Leucocytozoidae and Haemoproteidae.) *Can. J. Zool.* **43,** 927–932.

Berson, J. P. (1964). Les Protozoaires parasites des hematies et du systeme histiocytaire des oiseaux. Essai de nomenclature. *Revue Elev. Méd. vét. Pays trop.* **14,** 43–96.

Borg, K. (1953). On *Leucocytozoon* in Swedish Capercaillie, Black Grouse and Hazel Grouse. *Akad. Avh., Lund,* 1–109.

Box, Edith D. (1967). Influence of *Isopora* infections on patency of avian *Lankesterella* (*Atoxoplasma,* Garnham, 1950). *J. Parasit.* **53,** 1140–1147.

Chernin, E. (1952). The epizootiology of *Leucocytozoon simondi* infections in domestic ducks in northern Michigan. *Am. J. Hyg.* **56,** 39–57.

Coatney, G. R. (1936). A check-list and host-index of the genus *Haemoproteus.* *J. Parasit.* **22,** 88–105.

Coatney, G. R. (1937). A catalog and host-index of the genus *Leucocytozoon.* *J. Parasit.* **23,** 202–213.

Coatney, G. R. and Roudabush, R. L. (1936). A catalog and host-index of the genus *Plasmodium. J. Parasit.* **22,** 338–353.

Coatney, G. R. and Roudabush, R. L. (1949). A catalog of the species of the genus *Plasmodium* and index of their hosts. Boyd's Malariology, I: 29–53. W. B. Saunders Company, Philadelphia and London.

Corradetti, A., Garnham, P. C. C. and Laird, M. (1963). New classification of the avian malaria parasites. *Parassitologia* **5,** 1–4.

Corradetti, A., Neri, I. and Scanga, M. (1960). Segnalazione in Italia di *Plasmodium praecox* var. *matutinum* in *Turdus iliacus* e separazione di questo plasmodio da *Plasmodium praecox* come specie distinta; *Plasmodium matutinum* Huff. *Parassitologia* **2,** 333–343.

Danilewsky, B. (1884). On the parasites of the blood (Haematozoa). *Russk. Med.,* pp. 46, 48.

Diamond, L. S. and Herman, C. M. (1954). Incidence of trypanosomes in the Canada goose as revealed by bone marrow culture. *J. Parasit.* **40**, 195–202.

Fallis, A. M. and Bennett, G. F. (1960). Description of *Haemoproteus canachites* n. sp. (Sporozoa: Haemoproteidae) and sporogony in Culicoides (Diptera: Ceratopogonidae). *Can. J. Zool.* **38**, 455–464.

Fallis, A. M. and Bennett, G. F. (1966). On the epizootiology of infections caused by *Leucocytozoon simondi* in Algonquin Park, Canada. *Can. J. Zool.* **44**, 101–112.

Fallis, A. M. and Wood, D. M. (1957). Biting midges (Diptera: Ceratopogonidae) as intermediate hosts for *Haemoproteus* in ducks. *Can. J. Zool.* **35**, 425–435.

Fiennes, R. N. T-W. (1967). Penguin pathology. *Int. Zoo. Yb.* **7**, 11–14.

Garnham, P. C. C. (1950). Blood parasites of East African vertebrates, with a brief description of exo-erythrocytic schizogony in *Plasmodium pitmani*. *Parasit.* **40**, 328–337.

Garnham, P. C. C. (1966). Malaria parasites and other haemosporidia. Blackwell Scientific Publ., Oxford. 1114 pp.

Giovannola, A. (1939). I plasmodi aviara. *Riv. Parassit.* **3**, 221–266.

Hanson, H. C., Levine, N. D., Kossack, C. W., Kantor, S. and Stannard, L. J. (1957). Parasites of the mourning dove (*Zenaidura macroura carolinensis*) in Illinois. *J. Parasit.* **43**, 186–194.

Herman, C. M. (1937). *Toxoplasma* in North American birds, and attempted transmission to canaries and chickens. *Am. J. Hyg.* **25**, 303–312.

Herman, C. M. (1938). Epidemiology of malaria in eastern redwings (*Agelaius p. phoeniceus*). *Am. J. Hyg.* **28**, 232–243.

Herman, C. M. (1944). The blood protozoa of North American birds. *Bird-Banding* **15**, 80–112.

Herman, C. M. (1954). *Haemoproteus* infections in waterfowl. *Proc. helminth. Soc. Wash.* **21**, 37–42.

Herman, C. M. (1963). The occurrence of protozoan blood parasites in Anatidae. *Transactions of the VIth Congress, Int. Union Game Biol., The Nature Conservancy, London*, 341–349.

Herman, C. M. (1968). Blood parasites of North American Waterfowl. *Trans. N. Am. Wildl. Conf.* **33**, 348–359.

Herman, C. M. and Bischoff, A. I. (1949). The duration of *Haemoproteus* infection in California quail. *Calif. Fish Game* **35**, 293–299.

Herman, C. M., Knisley, J. O. and Snyder, E. L. (1966). Subinoculation as a technique in the diagnosis of avian *Plasmodium*. *Avian Dis.* **10**, 541–547.

Herman, C. M., Kocan, R. M., Snyder, E. L. and Knisley, J. O. (1968). *Plasmodium elongatum* from a penguin. *Bull. Wildl. Dis. Ass*, **4**, 132.

Herman, C. M., Reeves, W. C., McClure, H. E., French, E. M. and Hammon, W. McD. (1954). Studies on avian malaria in vectors and hosts of encephalitis in Kern County, California. I. Infection in avian hosts. *Am. J. trop. Med. Hyg.* **3**, 676–695.

Hewitt, R. I. (1940). Bird Malaria. Johns Hopkins University Press, Baltimore, 228 pp.

Huff, C. G. (1932). Studies on *Haemoproteus* of Mourning Doves. *Am. J. Hyg.* **16**, 618–623.

Knisley, J. O. and Herman, C. M. (1967). *Haemoproteus*, a blood parasite in domestic pigeons and mourning doves in Maryland. *Chesapeake Sci.* **8**, 200–205.

Knuth, P. and Magdeburg, F. (1922). Uber ein durch Leukozytozoen verursachtes Sterben junger Ganse. *Berl. tierarztl. Wschr.* **33**, 359–361.

Knuth, P. and Magdeburg, F. (1924). Ueber Leukozytozoen bei der Hausgans. *Z. InfecktKrankh. parasit. Krankh. Hyg. Haustiere* **26**, 42–52.

Kocan, R. M. and Clark, D. T. (1966). Anemia in Ducks Infected with *Leucocytozoon simondi*. *J. Protozool.* **13**, 465–468.

Lainson, R. (1959). *Atoxoplasma* Garnham, 1950, as a synonym for *Lankesterella* Labbé, 1899. Its life cycle in the English sparrow (*Passer domesticus domesticus*, Linn). *J. Protozool.* **6**, 360–371.

Laird, M. and Lari, F. A. (1957). The avian blood parasite *Babesia moshkovskii* (Schurenkove, 1938), with a record from *Corvus splendens* Veillot in Pakistan. *Can. J. Zool.* **35**, 783–795.

Levine, N. D. and Kantor, S. (1959). Check list of blood parasites of birds of the order Columbiformes. *Wildl. Dis.* No. 1, 1–38.

Lucena, D. T. (1939). Malaria aviaria. *Jornal do Commercio. Recife, Brazil*, 126 pp.

Manwell, R. D. (1935). How many species of avian malaria are there? *Am. J. trop. Med.* **15**, 365–380.

Manwell, R. D. (1938). The identification of the avian malarias. *Am. J. trop. Med.* **18**, 565–575.

Mathis, C. and Leger, M. (1910). Sur des *Haemoproteus* de quelques oiseaux du Tonkin. *Bull. Soc. Path. exot.* **3**, 704–708.

Novy, F. G. and MacNeal, W. J. (1904). Trypanosomes and bird malaria. *Am. Med.* **8**, 932–934.

O'Roke, E. C. (1934). A malaria-like disease of ducks. *Bull. Sch. For. Conserv. Univ. Mich.*, **4**, 1–44.

Purchase, H. S. (1942). Turkey malaria. *Parasit.* **34**, 278–283.

Ross, R. (1898). Report on the cultivation of *Proteosoma*, Labbe, in grey mosquitoes. *Indian med. Gaz.* **33**, 401–408, 448–451.

Stabler, R. M., Holt, P. A. and Kitzmiller, N. J. (1966). *Trypanosoma avium* in the blood and bone marrow of 677 Colorado birds. *J. Parasit.* **52**, 1141–1144.

Warner, R. E. (1968). The rôle of introduced diseases in the extinction of the endemic Hawaiian avifauna. *Condor* **70**, 101–120.

Wenyon, C. M. (1926). Protozoology—2 volumes. Ballière, Tindall and Cox, London, 1563 pp.

L

CHAIRMAN'S SUMMING-UP

Some interesting new ideas have been put forward in these papers and various vital lacunae in our knowledge have been stressed, particularly on the degree of pathogenicity of the infections in animals (young and old) in the wild. Mr. Wells and Dr. Lumsden stress the importance of the correct identification of trypanosomes and the need for new serological methods for detecting them and Dr. Carlton Herman emphasizes the same point in relation to protozoa in the blood of birds, which, however, can be revealed in far higher percentages by the use of special techniques, such as isodiagnosis and xenodiagnosis. Dr. Baker's recent work on wild animals in the Serengeti makes it fairly certain that trypanosomiasis in these animals is often acquired by the consumption of infected prey and possibly also by mechanical transmission. Dr. Barnett and Dr. Brocklesby presented a masterly survey of the piroplasms in the animal kingdom and took *Theileria parva* and *T. lawrencei* as an example of the transference of these organisms from a wild to a domestic source and vice versa; they also pointed out how *Cytauxzoon* species are pathogenic to their natural hosts, unlike the general behaviour of protozoa in wild animals.

ECTOPARASITES, ENDOPARASITES AND MYCOSES

CHAIRMAN: K. C. SELLERS

CHAIRMAN'S INTRODUCTION

In preparing to chair this session I browsed through a number of text books dealing with ecto- and endoparasites and mycoses. In one of these standard text books I came across a statement that *Oestridae* flies can reach 800 m.p.h. in flight; perhaps a collision between a warble fly and a cow might have calamitous results. However, I checked this statement with an active entomologist who assured me that this was another of these inherited text book errors and *Oestridae*, in fact, had difficulty in attaining 40 m.p.h. all out.

The speakers this afternoon have not, I can assure you, depended upon text books for their information and between them they have obviously achieved a great deal of work in their various fields. Firstly, Dr. Blackmore and Miss Dawn Owen from the Medical Research Council will cover ectoparasites of British wild rodents; secondly, Dr. Dunn from the Veterinary School at the University of Glasgow will be dealing with endoparasites in wild ruminants and finally Mr. Austwick from the Central Veterinary Laboratory at Weybridge will be discussing various mycological problems. Under the circumstances, I have decided to defer discussion of the individual papers until the end of the session as the length of the papers may vary somewhat and questions and comment will undoubtedly be affected by their contents.

Symp. zool. Soc. Lond. (1968) No. 24, 197–220.

ECTOPARASITES: THE SIGNIFICANCE IN BRITISH WILD RODENTS

D. K. BLACKMORE and D. G. OWEN

*Laboratory Animals Centre, M.R.C. Laboratories,
Carshalton, Surrey, England*

SYNOPSIS

Some of the more general aspects of the significance of ectoparasitism of free-living wild animals are discussed. The information available concerning such infestation in British wild rodents is reviewed, with special reference to the ticks and mites (Acarina), fleas (Siphonaptera) and lice (Phthiraptera) and to the host specificity of each group of parasites.

The results of a small survey of the ectoparasites of both free-living and captive brown rats (*Rattus norvegicus*) and house mice (*Mus musculus*) is recorded, together with comments on the effects of the environment on the degree and type of infestation.

Ectoparasitism of laboratory rodents in Britain is reviewed, together with the results of a survey of such infestation of 612 smaller laboratory rodents obtained from 64 different sources.

The general lack of information concerning the ectoparasites of free living wild rodents and some of the more important implications of ectoparasitism of wild rodents is discussed. Apart from the general ecological aspects of ectoparasitism, the significance of such infestation in relation to the transmission of disease to domesticated stock is discussed, particular emphasis being paid to the hazard to laboratory animals.

INTRODUCTION

The majority of mammalian ectoparasites belong to the classes Arachnida and Insecta. The former class contains the mites and ticks, while the latter contains the fleas, lice and parasitic flies. Ectoparasites are extremely well adapted to their way of life and almost any free-living wild animal is permanently infested with a variety of species. Some infestations, particularly by blood sucking parasites, tend to be seasonal in their incidence. The type and numbers found depend on a variety of factors, the most important of which are the species of animal concerned, its geographical distribution and its general state of health. As it would be impossible to attempt to cover the complete subject of ectoparasitism of free-living wild animals in the time available, it has been decided to consider only the ectoparasites of British wild rodents and to discuss certain implications of such infestations.

British wild rodents are confined to four families, the Gliridae, the Muridae, the Sciuridae and the Capromyidae. The only representative of the Gliridae are the Dormice (*Muscardinus avellanarius* and *Glis glis*). The Muridae are represented by the voles (*Clethrionomys glareolus, Microtus agrestis, Arvicola amphibius*), the field mouse (*Apodemus sylvaticus*), the harvest mouse (*Micromys minutus*), the house mouse (*Mus musculus*), the brown rat (*Rattus norvegicus*) and the black rat (*Rattus rattus*). The grey squirrel (*Sciurus carolinensis*) and the red squirrel (*Sciurus vulgaris*) comprise the family Sciuridae. The coypu (*Myocaster coypus*) is the only member of the Capromyidae.

The actual significance of any disease of free-living wild animals is usually considered from one of two basic points of view: firstly, the effect of the disease on the animal itself and the resultant ecological implications and secondly, from the relationship of the disease in the wild species and its infectivity for man or his domesticated animals.

Although ectoparasitism of the skin and gills of fish by *Gyrodactylus elegans* (Van Duijn, 1956) and heavy infestation of young birds with blood sucking arthropods can cause death, ectoparasites are seldom a primary or sole cause of death. Their pathological significance is therefore more often associated with their capability to act as vectors of other microbial or parasitic diseases, obvious examples being the mosquito as a vector of malaria, yellow fever and many other diseases, tsetse flies transmitting trypanosomiasis, the rat flea (*Xenopsylla cheopis*) as a vector of bubonic plague and the rabbit flea (*Spilopsyllus cuniculi*) as one of the vectors of myxomatosis.

Information concerning ectoparasites as disease vectors is comparatively well documented and the general direct and indirect effects of parasites on their hosts have been well summarized by Sprent (1963).

Although certain groups of ectoparasites of free-living wild animals such as the fleas (Rothschild, 1915; Smit, 1957a, 1957b) and ticks (Arthur, 1963) have been discussed, information concerning the whole spectrum of ectoparasites is somewhat sparse. The information available will be reviewed, together with the results of a small survey of the ectoparasites of brown rats carried out by one of the authors (D. O.). Before discussing the significance of this naturally occurring parasitic burden, the more common parasites of laboratory animals will also be reviewed.

Over the decade, almost all research workers have come to realize the importance of using a standard disease-free animal for research purposes (Bleby, 1967). With the resultant increasing need for S.P.F. (specific pathogen free) animals, much greater precautions must be taken to avoid infection with any pathogen, whether microbial or parasitic, and free-living wild rodents can become a very real hazard to the health of a laboratory animal colony.

It has been decided in this contribution to compare the ectoparasites of British wild rodents with those encountered in the smaller laboratory animals.

THE ECTOPARASITES OF WILD RODENTS

Ticks (ixodoidea)

Ticks are a group of parasites which do not have a rigid host specificity but of the 21 species of true British ticks there are records of only 6 which are potentially capable of affecting our wild rodent population (Arthur, 1963).

The two species of tick most commonly found on wild rodents, and with the widest distribution, are *Ixodes ricinus* and *Ixodes trianguliceps*. Immature stages of *I. ricinus* are capable of affecting a wide variety of species of mammals and birds and infestation has been recorded in 29 species of mammal and 39 species of bird in Britain (Milne, 1949).

Although adult and immature stages of the tick are well recognized and important parasites of farm stock, immature forms have been recorded on short-tailed voles, field mice, red squirrels (Milne, 1949) and the coypu (Davis and Shillito, 1967). *I. trianguliceps* is essentially a rodent tick and all stages of the parasite have been found on voles (including the water vole), field mice, house mice and both species of rat. Both these ticks show definite seasonal incidence and peak infestations occur in the spring and autumn. In the case of *I. trianguliceps*, up to 70% of wild rodents may be affected during these peak periods, while in the winter the incidence is less than 20% (Cotton and Watts, 1967). Larvae and nymphs are found mainly around the ears, while the adults are found on the neck and body.

I. hexagonus, which is primarily a parasite of hedgehogs, has also been recorded from the brown rat (Arthur, 1963) and the coypu (Davis and Shillito, 1967), while *I. arivolae* has only been reported on water voles and coypu. *Dermacentor reticulatus* is essentially a tick of larger farm animals, mainly confined to the South-West regions of the country, but larvae and nymphs have been recorded from short-tailed voles and water voles. *I. festai*, although not yet actually recorded from any British wild rodent and confined to Lundy Island, is known elsewhere in the world to be capable of affecting field mice, house mice and the black rat.

Mites (acarina)

Both parasitic and non-parasitic mites are extremely numerous and a very large proportion of the soil fauna consists of various non-parasitic species (P. Murphy, 1968, personal communication). In spite of the huge naturally occurring mite population, information concerning the general ecology and host associations of the parasitic mites of British wild animals is extremely sparse. A certian amount of information on their identification has been published (Hirst, 1915, 1916, 1919), together with records of the isolation of particular species of mites from rather small samples of wild rodents in widely differing geographical areas. One of the few contributions relevant to a review of the mites of wild rodents was published in 1934 and lists 16 species recovered from house mice and field mice on the Isle of Lewis in the Outer Hebrides (Hora, 1934).

Mites found in association with any wild animal are either obligatory parasites or facultative parasites that are primarily nest dwellers or free-living mites which are inadvertently found on the animal. This differentiation between parasitic and free-living mites has been discussed in relation to the Dermanyssidae by Evans and Till (1966). Of the 16 mites recovered from mice by Hora (1934), only 11 were considered to be obligatory parasites.

Obligatory parasites can be divided into four main groups: the skin burrowing and hair follicle mites, the surface feeders, the nest dwellers which only attack the host to feed and, finally, those which only parasitise the host during a particular stage of their life cycle.

Notoedres notoedres (see Fig. 1) is a typical example of a burrowing mite and it appears to occur relatively frequently on the brown rat (Balfour, 1922). Two species of the follicular mite, *Demodex ratti* and *D. nanus,* have also been recorded from the rat (Hirst, 1919), while *Psorergates simplex* has been recorded from laboratory mice (Cook, 1956; Beresford-Jones, 1965).

There are many examples of the surface dwelling mites which feed on either tissue fluid or epidermal tissue. Both *Myobia musculi* (see Fig. 2) and *Mycoptes musculinus* (see Fig. 3) frequently occur on both brown rats and house mice.

A large number of mites belong to the group which are nest dwellers but attack their host to feed and the majority of this group belong to the Dermanyssidae. *Laelaps echidninus* occurs in both species of rat (Hirst, 1915; Balfour, 1922) and the field mouse (Thomson, 1935) and *Eulaelaps strabularis* is found on rats (Hirst, 1915) and both the house mouse and field mouse (Hora, 1934; Thomson, 1935). There are many records of related species occurring on other British wild rodents.

The harvest mite (*Trombicula autumnalis*) is the best known example of a mite, the larvae of which are the only parasitic stage (Richards, 1950; Radford, 1954). This mite is apparently capable of attacking any species of mammal and some species of bird. Due to the fact that only larval stages are parasitic, there is a definite seasonal incidence of infestation. The main brood larvae appear towards the end of June and

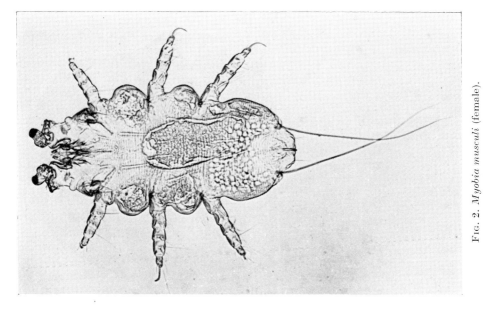

Fig. 2. *Myobia musculi* (female).

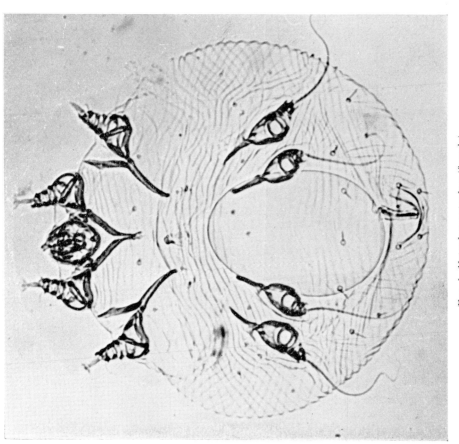

Fig. 1. *Notoedres notoedres* (female).

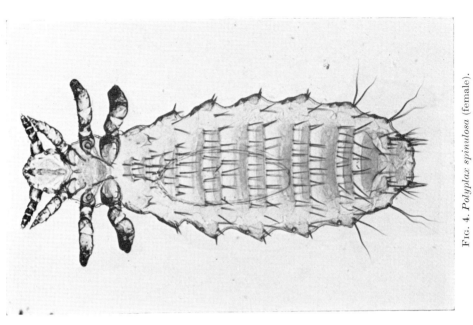

Fig. 4. *Polyplax spinulosa* (female).

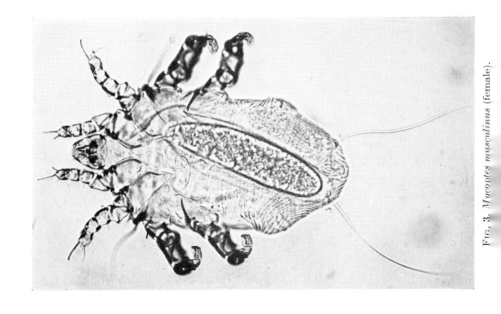

Fig. 3. *Mucoptes masculinus* (female).

numbers reach a maximum in September (Elton and Keay, 1936; Richards, 1950). However, Elton and Keay (1936) found that field voles and field mice were only infected during the autumn.

Space does not permit the inclusion of the many records of different and often obscure mites which have been isolated from British wild rodents. In general principle the parasitic mites do not have a rigid host specificity and it is probable that in many instances wild rodents are capable of being infested with a similar spectrum of parasitic mites.

Fleas (siphonaptera)

The fleas are somewhat similar to ticks in that adult stages of the parasite do not exhibit a rigid host specificity. Of the 56 species of British flea, at least 21 are known to affect free-living wild rodents (Smit, 1957a) and the majority of these 21 species occur comparatively frequently. A list of these fleas, together with a summary of their hosts and geographical distribution is shown in Table I. It is interesting to note that the coypu does not appear to suffer from any primary flea infestation (Davis and Shillito, 1967) although the chicken flea (*Ceratophyllus gallinae*) has been recorded from this rodent (George, 1964).

Although adult fleas do not exhibit a rigid host specificity, their immature free-living forms are more adapted to a particular environment and larval stages are much more nest specific than the adults are host specific. These circumstances result in the fact that the primary fleas of rodents with similar nesting or breeding habitats are of similar or identical species. Both the red and grey squirrels, which have a similar arboreal habitat, are affected with similar species of fleas and these fleas are not commonly found on the ground nesting rodents. *Orchopeas howardi*, which was introduced to this country with the grey squirrel at the beginning of the century, now commonly infests the red squirrel. Conversely, the common primary fleas of the urban rat and house mouse are dissimilar from the primary fleas of the voles and field mice which breed in a very different rural environment.

Temporary infestation of a rodent by a secondary flea from a species of animal with a different breeding habitat is occasionally encountered. Grey squirrels have been found to be infested with *Spilopsyllus cuniculi*, the common rabbit flea (Freeman, 1941), and rats with cat fleas (*Ctenocephalides felis*) (Balfour, 1922) and dog fleas (*Ctenocephalides canis*) (Newstead and Evans, 1921). Such unexpected infestations are presumably due to cross infection by adult stages of the flea, rather than from fleas which have developed from immature stages infesting the nest of the host.

TABLE I

Flea infestation of British wild rodents

Species of flea	Primary host	Secondary host	Distribution in U. K.	Frequency of occurrance
Xenopsylla cheopis	Black rat Brown rat	—	In vicinity of ports	Uncommon, except in vicinity of ports
Xenopsylla brasiliensis	Black rat Brown rat	—	In vicinity of ports	Very rare
Nosopsyllus fasciatus	Black rat Brown rat	House mouse Field mouse Field vole	Throughout U.K.	Common
Nosopsyllus londiniensis	Black rat House mouse	Field mouse	In vicinity of ports	Common locally
Monopsyllus sciurorum	Red squirrel Grey squirrel Dormice	Voles	Throughout U.K.	Common
Malaraeus penicilliger mustelae	Bank vole Field vole	Field mouse Dormouse	Throughout U.K.	Common
Megabothris walkeri	Bank vole Field vole Water vole	—	Throughout U.K.	Fairly common
Megabothris turbidis	Bank vole Field mouse	—	England and Wales	Common
Megabothris rectangulatus	Bank vole Field vole	—	Scotland	Common

Ctenophthalmus nobilis nobilis	Brown rat House mouse Field mouse Voles	—	S.W. England and coastal areas	Common
Ctenophthalmus nobilis vulgaris	As above	—	Throughout U.K.	Common
Euctenophthalmus conger conger	Bank vole	—	S.E. and E. England	Rare
Rhadinopsylla pentacanthia	Bank vole Field vole Field mouse	—	Throughout U.K.	Fairly common
Rhadinopsylla iscanthia	Bank vole	—	England	Rare
Rhadinopsylla integella	Bank vole Field vole	—	Scotland	Uncommon
Leptopsylla segnis (musculi)	House mouse	Field mouse	Throughout U.K.	Fairly common
Perdmyscopsylla silvatica spectabilis	Bank vole Field vole	—	Throughout U.K.	Uncommon
Typhloceras poppei	Field mouse	House mouse	Throughout U.K.	Fairly common
Hystrichopsylla talpae	Mole Bank vole Shrews	Field mouse Field vole	Throughout U.K.	Common
Taropsylla octodecimdentata octodecimdentata	Red squirrel	—	Scotland	Fairly common
Orchopeas howardi	Grey squirrel Red squirrel Edible dormouse	—	Throughout U.K.	Common

Bacot (1919) listed 18 species of flea recovered from rats in this country but only considered the rat to be a true or primary host for 7 or 8 of the species.

The number of fleas found on an animal is subject to considerable variation. One factor responsible for this variation is associated with the time the animal spends in the nest, where the immature stages of the parasite develop. Freeman (1941) showed that $64\cdot3\%$ of adult breeding grey squirrels carried fleas, compared with only $31\cdot1\%$ of non-breeding animals. This latter group of animals spend far less time in a permanent "drey".

Lice (phthiraptera)

Although Hopkins (1949) has published a general review of the host associations of the lice of mammals, other information concerning louse infestation of rodents is extremely sparse. A total of only 15 species of lice are recorded as affecting this group of animals (see Table II) and it is interesting to note that 14 belong to the sub-order Anoplura and only one to the Mallophaga, *Pitrufquenia coypus*, which is specific to the coypu (established as a British feral animal in 1942) and which was introduced from South America with it (Freeman, 1946).

Unlike ticks and fleas, lice have a fairly rigid host specificity although closely related species of rodent may be infected with the same species of louse. The red and grey squirrel can be hosts to *Neohaematopinus sciurinus*, the brown and black rat are infested with *Polyplax spinulosa* (see Fig. 4, p. 204) and both the field mouse and house mouse can be affected by *Polyplax serrata*.

Hopkins (1949) stated that rodents usually have rather light infestations and seldom carry more than 20 lice. However, this statement does not agree with our own findings in relation to the brown rat and Hopkins himself refers to a record of 6000 specimens on a rat (Eichler, 1940).

As lice can only survive off their host under normal conditions for a few hours (Lapage, 1956), infection between rodents must be almost entirely due to direct contact.

Heavy louse infestation by sucking lice can be of direct significance to the host. Polyplax is known to transmit tularaemia and murine typhus and has been incriminated in the spread of Haemobartonella and Epyrithrozoon. Owing to their rather rigid host specificity, they are unlikely to be the direct cause of the spread of such diseases between different species of wild rodents although they may be an important factor in disseminating a microbial systemic disease amongst a particular community.

TABLE II

Louse infestation of British wild rodents

Parasite	Host
Enderleinellus nitzchi	Red squirrel
Enderleinellus longiceps	Grey squirrel
Neohaematopinus sciurinus	Grey squirrel
	Red squirrel
Hoplopleura scuricola	Grey squirrel
Hoplopleura longula	Harvest mouse
Hoplopleura affinis	Field mouse
Hoplopleura pacifica	Black rat
Hoplopleura hesperomydis	House mouse
Hoplopleura acanthopus edentatus	Bank vole
Hoplopleura acanthopus acanthopus	Field vole
Polyplax serrata	Field mouse
	House mouse
Polyplax spinulosa	Black rat
	Brown rat
Polyplax gracilis	Harvest mouse
Polyplax spinigera	Water vole
Schizophitherous pleurophaeus	Dormouse
Pitrufquenia coypus	Coypu

Observations on the ectoparasites of brown rats and house mice

One of the authors (D.O.) made a limited study of the ectoparasites of free-living and captive brown rats and house mice. Although relatively few animals were examined, these results are included because so little general information on this subject is available.

Thirty rats from the Liverpool area were examined during an 8-month period. The animals were trapped from bombsites, derelict buildings and sewers and were killed and examined within 25–48 h of capture. Fifteen rats were heavily infested with the "fur mite" *Myobia musculi*, 10 with the "ear mange mite" *Notoedres notoedres* and 6 were affected with both species of mite. In many cases, the infestation due to *Notoedres notoedres* was severe in animals of all ages also affecting the scrotum, outer ear, feet and base of the tail. Twenty-three (77%) of the rats were affected by the blood sucking louse *Polyplax spinulosa*. These infestations were most severe in older animals or those suffering from concurrent disease. Only 4 rats were affected with fleas, the infestations were light and were due to *Ctenopthalmus noblis noblis* and *Nosopsyllus fasciatus*. It is interesting to note that only one rat in this series was infected by *Trypanosoma lewisi* which requires a flea as vector. Six mice inadvertently caught in the rat traps were infected with *Myobia musculi*, *Mycoptes musculinus* and *Polyplax serrata* but no fleas were found.

More recently, a small number of brown rats and house mice were examined from the vicinity of the sheep and goat paddocks at Carshalton. The parasitic burden was similar to that of the animals from Liverpool, except that larvae of *Ixodes* sp. were found during the spring and autumn.

Twenty-four rats and 48 house mice were examined from the captive colony maintained by the Infestation Control Laboratories of the Ministry of Agriculture, Fisheries and Food at Tolworth. The majority of these animals were wild caught and had been in captivity for varying periods. The rats had been maintained on wire bottom cages over metal trays, which had been regularly cleaned and disinfected. The only parasites found on these rats consisted entirely of those whose complete life cycle is spent on the animals. They were lightly infested with *Myobia musculi* and heavily infested with *Polyplax spinulosa*. The mice had been kept in batches of approximately 50 in large bins, the floors of which were covered with sawdust and hay. All the animals were lightly infested with *Myobia musculi* and/or *Mycoptes musculinus*, and/or *Polyplax spinulosa*, and without exception with a heavy flea infestation of 7–15 fleas per animal due to *Ctenophthalmus noblis noblis*. The majority of the mice were in addition heavily infested by the cestode *Hymenolepis microstoma* which infests the bile ducts and duodenum (see Fig. 5) and requires an arthropod intermediate host. Fleas are well recognized vectors of disease and the large number of cestodes recorded in these mice may be associated with their heavy infestation of fleas.

ECTOPARASITES OF LABORATORY RODENTS

The more common laboratory rodents are rats, mice, guinea-pigs and hamsters. The laboratory rat has been developed from the brown rat (*Rattus norvegicus*) (Porter, 1967) and the laboratory mouse from the house mouse (*Mus musculus*) (Staats, 1966). Guinea-pigs (*Cavia porcellus*) were developed from the free-living South American guinea-pig (*Cavia cutleri*) (Paterson, 1967), while the Syrian or golden hamsters (*Mesocricetus auratus*) are direct descendants of the animals found in Syria in 1932 (Magalhaes, 1967). Chinese hamsters (*Cricetulus griseus*) are still found in the wild in the area between the Caspian sea and the east coast of China (Whitney, 1963).

The degree of ectoparasitism of laboratory rodents is completely dependent on the standard of husbandry and hygiene to which the animals are subjected. No ectoparasites, therefore, will be found affecting animals which are completely barrier maintained, such as those termed

FIG. 6. *Chirodiscoides caviae* (female).

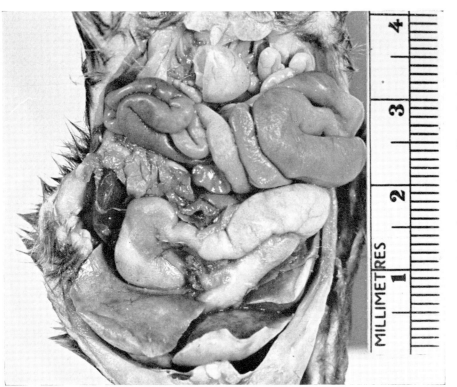

FIG. 5. Enlarged bile ducts of mouse due to *Hymenolepis microstoma*.

212 D. K. BLACKMORE and D. G. OWEN

S.P.F. or germfree. Although textbooks (UFAW, 1967; Oldham, 1967; Tuffery and Innes, 1963) refer to ectoparasites of laboratory animals, no exact information on the incidence of such infestation is given or the infestations described related to laboratory animals in other countries. A survey was carried out, therefore, on parasitic infestation of animals supplied by approved breeders in this country to research establishments. Six hundred and twelve animals from 64 different breeders were examined, consisting of 394 mice, 48 rats, 80 guinea-pigs, 36 hamsters and 34 rabbits. Unfortunately, from the point of view of this contribution, all 48 rats were from barrier maintained colonies and were completely free of all ectoparasites and it is beyond the scope of this paper to discuss the parasites recorded from the rabbits.

Ticks and fleas (ixodoidea and siphonaptera)

No ticks or fleas were isolated from any of the animals received. This was not unexpected as the normal husbandry of conventional laboratory animal colonies would prevent the animals from coming into contact with the free-living stage of these parasites. However, Oldham (1967) suggests that 4 species of flea (*Leptopsylla segnis*, *Nosopsyllus fasciatus*, *Ctenophthalamus agyrtes* and *Xenopsylla cheopis*) can occur in laboratory rats and mice.

Mites (acarina)

The 394 mice examined were from 13 different colonies, 10 of which were found to be infested with *Myobia musculi* and 4 concurrently with *Mycoptes musculinus*. Only a few of the older animals showed macroscopic evidence of infection.

Although no other species of mite were isolated, *Psorergates simplex* has been recorded from laboratory mice in this country (Cook, 1956; Beresford-Jones, 1965).

As already stated, no mites were isolated from any of the rats but *Myobia musculi* has frequently been isolated from rats not included in this survey.

The 80 guinea-pigs were from 20 different colonies, 18 of which were found to be heavily infested with *Chirodiscoides caviae* (see Fig. 6, p. 211).

The authors have not isolated mites or other ectoparasites from hamsters, even from rather overcrowded colonies, and it is interesting to note that although Hindle (1949) records *Notoedres notoedres* (see Fig. 1) as commonly occurring in this species, Wantland (1955) in America was unable to support this statement from any of the hamster colonies which he examined. However Fulton (1943) has also recorded this infestation in hamsters. Nutting (1961) recorded the isolation of two different species of Demodex from hamsters (*D. aurati* and *D. criceti*).

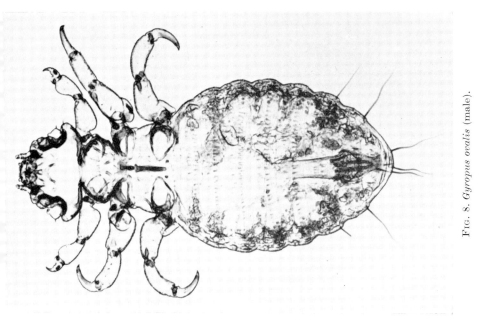

Fig. 8. *Gyropus ovalis* (male).

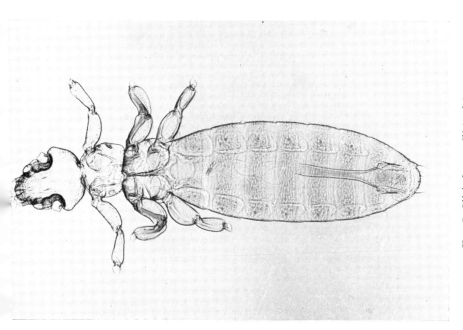

Fig. 7. *Gliricola porcelli* (male).

Oldham (1967) lists 7 species of mite as parasites of laboratory rats and mice but it appears that this is a list of mites which are potentially capable of causing infestations, rather than those which have been recorded as actually causing infection in this country.

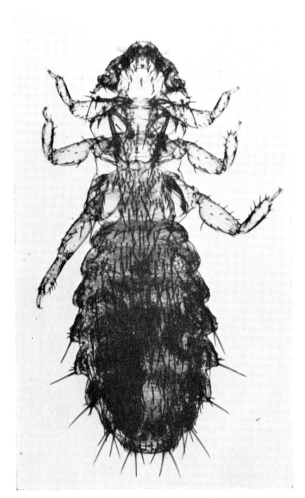

Fɪɢ. 9. *Trimenopon jenningsi* (female).

Lice (*phthiraptera*)

Four of 13 mouse colonies were infested with *Polyplax serrata* but the animals were only lightly infected. Although none of the rats in this survey were found to have lice, *Polyplax spinulosa* has been quite frequently recovered from other colonies.

Of the 20 guinea-pig colonies examined, 19 were infested with *Gliricola porcelli* (see Fig. 7). Eighteen were concurrently infested with *Gyropus ovalis* (see Fig. 8) and 6 with *Trimenopon jenningsi* (see Fig. 9). Infestations were frequently heavy and caused little or no obvious clinical effects except in older animals or those in poor condition which showed evidence of slight loss of hair and coarsening of the general appearance of the coat.

The various species of louse recorded in this small survey were similar to those listed in the majority of text books as the usual lice that cause infestation in laboratory rodents.

No other types of ectoparasites have been recovered from any of the laboratory rodents examined but both Porter (1967) and Tuffery and Innes (1963) mention the bed bug (*Cimex lectuloris*) as a possible parasite of rats and mice and Wantland (1955) recorded one case of rectal myasis due to the larvae of *Musca domesticus* in a hamster.

DISCUSSION

A discussion of the ectoparasites of wild rodents can be conveniently divided into 3 main parts. Firstly, a consideration of the parasites themselves; secondly, their effect on their host and, finally, the implications of rodent infestation on man and his domesticated stock. Although there is a reasonable amount of information on the general morphology and identification of ectoparasites, there are few facts on their general ecology with the possible exception of the ticks. Probably the major reason for this dearth of information is the previous lack of interest in almost all aspects of the diseases of wildlife by the major British research organizations. The collection of information on ectoparasites, therefore, has been left almost entirely to enthusiastic individuals, the majority of whom have been entomologists. Before the true significance of the ectoparasites of wild rodents can be appreciated, much more information is required, needing closer team work and facilities not at present available.

The varied degree of host specificity of ectoparasites is of importance. At one end of the scale are the lice, which show a rather rigid host preference, while at the other are certain species of mites, which can infest almost any mammal. The ticks are much less host specific than the lice. *Ixodes trianguliceps* is essentially a rodent tick and, although *Ixodes ricinus* is primarily a cattle tick, immature stages are capable of feeding on most species of mammal. The larvae of fleas are much more sensitive to their micro-environment than are the adult stages and so rodents with similar nesting environments tend to be affected by a similar spectrum of fleas.

The geographical distribution of ectoparasites is dependent on several factors, the most important of which are the distribution of the host and climatic conditions. The published distribution can be confusing, as most surveys cover restricted areas, and in many parts of the country there are no records of the ectoparasites of the rodent population. Parasites which are strictly host specific, such as the coypu louse (*Pitrufquenia coypus*), are only found in the areas inhabited by their host. This louse also illustrates how a parasite can be introduced by allowing the importation of a new species of animal. The flea *Orchopeas howardi* was introduced to this country with the grey squirrel and is now widely distributed, affecting both species of squirrel. Parasites with free-living stages are influenced by the distribution of their primary hosts, for example rodents will not carry *Ixodes ricinus* if there are no cattle in the area. They are also influenced by climatic conditions, including seasonal variations, an example being harvest mite (*Trombicula autumnalis*) infestations which are only found in the summer months. Rodents, during the breeding season, spend more time in the nest and tend to carry a larger number of fleas. This latter point was well illustrated by Freeman (1941) in his study of grey squirrels and George and Corbet's (1959) observations that older voles, which inhabitated well established nests, carried a higher flea burden than younger animals living in newly constructed nests.

The skin burrowing mites are almost the only group of ectoparasites which have any directly adverse effect on their hosts. In both free-living and laboratory mice *Notoedres notoedres* can cause clinical skin lesions and Elton, Ford, Baker and Gardner (1931) recorded that 11% of the field mice examined had lesions on the legs associated with this parasite.

The most important effect of ectoparasites on rodents is as vectors of disease. They can act as intermediate hosts for various cestodes, examples being the relationship between fleas and *Hymolepis nana* and the infestation of captive house mice with *Hymolepis microstoma* referred to in this paper. Ectoparasites can also transmit various protozoan and other microbial diseases and the association of the flea and *Trypanosoma lewisi* is well established.

The parasites which are not particularly host specific may also be important in assisting the transmission of disease from one species of free-living animal to another. Elton *et al.* (1931) emphasized the importance of the fleas of wild mice as potential disseminators of disease.

More recently, the spread of myxomatosis has been definitely associated with ectoparasites.

The distribution on the host is often of interest; ticks are frequently confined to the head region, which may be partially due to the inability of the animal to remove them with its teeth, although immature stages may have a different distribution from the adults. Mites often have a predilection for certain areas of the body and in laboratory rodents *Notoedres notoedres* is usually found in the ears. The distribution of the three species of human louse are well known and the authors have noted a regional distribution of the three common lice affecting guinea-pigs: *Trimenopon jenningsi* occurs mainly on the rump, *Gliricola porcelli* on the back and flanks and *Gyropus ovalis* on the head and neck.

When a dual infestation occurs, one parasite may affect the degree of infestation of the other. In laboratory mice it has been noted that *Mycoptes musculinus* tends to limit the numbers of *Myobia musculi*. *Chyletiella parasitivorax* may also have an effect on other mite infestations owing to its predatory activities.

The relationship between the ectoparasites of wild rodents and domesticated animals is of major importance. Rodents can act as a reservoir of parasites which can directly affect man or his stock. Parasites which can survive on both wild rodents and domesticated stock are therefore capable of transmitting disease carried by the rodent population. The hazard of such disease transmission is less in this country than in other parts of the world but the high incidence of leptospirosis in rodents should not be forgotten, nor the possibility of the transmission of certain virus diseases.

Although wild rodents are not the most important source of disease for man and his domesticated animals, laboratory rodents are at considerable risk. As the laboratory mouse and rat have been evolved from the house mouse and brown rat, respectively, they are capable of being affected by the ectoparasites of their wild relations, whether or not the parasites are host specific. If such cross infection of ectoparasites occurs, the chance of disease transmission to a highly susceptible population is quite considerable. Apart from the laboratory rat and mouse, the other laboratory rodents, such as the guinea-pig and hamster, are susceptible to the less host specific parasites. Under normal laboratory management ticks are unlikely to be a hazard.

The majority of the preceding remarks have been theoretical assumptions, based on the rather limited information available. It is considered that this subject warrants much more study than has hitherto been carried out and, apart from the importance of ectoparasites of free-living wild animals to man and his farm stock, the direct and indirect hazards of wild rodent ectoparasites for laboratory animals should not be underestimated. Without disease-free experimental animals, much valuable research time and facilities are wasted.

ACKNOWLEDGEMENTS

The authors wish to thank the many people who have given help and advice in the collection of data for this paper. They would especially like to thank Dr. T. Hughes, Dr. A. Mead Briggs, Dr. P. Murphy, Mr. R. S. George and Mrs. M. Vizozo. Their gratitude is also expressed to Mr. R. Legg for the preparation of the photographic material.

REFERENCES

Arthur, D. R. (1963). British ticks. Butterworths, London.
Bacot, A. W. (1919). The fleas found on rats and their relation to plague. *Jl R. sanit. Inst.* **40**, 53.
Balfour, A. (1922). Observations on wild rats in England with an account of their ecto- and endoparasites. *Parasitology.* **14**, 282–298.
Beresford-Jones, W. P. (1965). Occurrence of the mite *Psorergates simplex* in mice. *Aust. vet. J.* **41**, 289–290.
Bleby, J. (1967). Specific-pathogen-free animals. *In* The U.F.A.W. Handbook on the care and management of Laboratory Animals. pp. 201–215, E. and S. Livingstone Ltd., Edinburgh and London.
Cook, R. (1956). Murine ear mange: The control of *Psorergates simplex* infestation. *Br. vet. J.* **112**, 22–25.
Cotton, M. J. and Watts, C. H. S. (1967). The ecology of the tick *Ixodes trianguliceps* Birula (Arachnida, Acarina, Ixodoidea). *Parasitology* **57**, 525–531.
Davis, R. A. and Shillito, E. (1967). The Coypu or Nutria. *In* The U.F.A.W. Handbook on the care and management of Laboratory Animals. pp. 457–467, E. and S. Livingstone Ltd., Edinburgh and London.
Eichler, W. (1940). *Arch. wiss. prakt. Tierheilk.* **75**, 212–221, quoted by Hopkins, G. H. E. (1949). *Proc. zool. Soc. Lond.* **119**, 1–4, 408.
Elton, C., Ford, E. B., Baker, J. R. and Gardner, A. D. (1931). Health and parasites of a wild mouse population. *Proc. zool. Soc. Lond.* **3**, 657–721.
Elton, C. and Keay, G. (1936). The seasonal occurrence of harvest mites (*Trombicula autumnalis* Shaw) on voles and mice near Oxford. *Parasitology.* **28**, 110.
Evans, O. G. and Till, W. M. (1966). Studies on the British Dermanyssidae (Acari: Mesostigmenta). Part I. External morphology. *Bull. Br. Mus. nat. Hist.* **13**, 15.
Freeman, R. B. (1941). The Distribution of *Orchopeas wickhami* (Baker) (Siphonaptera) in relation to its host the American Grey Squirrel. *Entomologist's mon. Mag.* **77**, 82–87.
Freeman, R. B. (1946). *Pitrufquenia coypus* Morelli (Mallophaga, Gyropidae) an ectoparasite on *Myocastor coypus* Mol. *Entomologist's mon. Mag.* **82**, 226–227.
Fulton, J. D. (1943). The treatment of Notoedres infections in golden hamsters (*Cricetus auratus*) with di-methyl-dyphenylene-disulphide ("Mitigal") and tetraethylthiuranimono-sulphide. *Vet. Rec.* **55**, 219.

George, R. S. and Corbet, G. B. (1959). A collection of fleas (*Siphonaptera*) from small mammals in the Scottish highlands. *Entomologist's Gaz.* **10**, 147–158.

George, R. S. (1964). *Ceratophyllus g. gallinae* (Schrank) from British coypu. *Entomologist's Gaz.* **15**, 40–41.

Hindle, E. (1949). The Golden Hamster Ch. 15, pp. 196–202 *In* The U.F.A.W. Handbook on the care and management of Laboratory Animals (A. N. Worden, ed.) Baillière, Tindall and Cox, London.

Hirst, S. (1915). Preliminary list of the Acari occurring on the brown rat (Rattus norvegicus) in Great Britain and the description of a new species (*Haemogomasus oudemonsi*). *Bull. ent. Res.* **5**, 119–124.

Hirst, S. (1916). Notes on the parasitic Acari. *J. zool. Res.* **1**, 59–81.

Hirst, S. (1919). Studies on the Acari I: The genus Demodex (Owen). British Museum, London.

Hopkins, G. H. E. (1949). Host associates of the lice of mammals. *Proc. zool. Soc. Lond.* **119**, 1–4, 387–604.

Hora, A. M. (1934). Notes on mites collected from the Isle of Lewis, Outer Hebrides. *Parasitology* **26**, 361–365.

Lapage, G. (1956). Veterinary Parasitology. Oliver and Boyd, Edinburgh and London.

Magalhaes, H. (1967). The Golden Hamster. *In* The U.F.A.W. Handbook on the care and management of Laboratory Animals, pp. 327–339, E. and S. Livingstone Ltd., Edinburgh and London.

Milne, A. (1949). The ecology of the sheep tick *Ixodes ricinus*. Host relationships of the tick, Pt. II, Observations on hill and moorland grazings in Northern England. *Parisitology*, **39**, 173–197.

Newstead, R. and Evans, A. M. (1921). Report of rat flea investigation. *Ann. trop. Med. Parasit.* **15**, 287.

Nutting, W. B. (1961). *Demodex aurati* and *D. criceli*; ectoparasites of the golden hamster (*Mesocricetus auratus*). *Parasitology*, **51**, 515–522.

Oldham, J. N. (1967). Helminths, Ectoparasites and Protozoa in Rats and Mice. *In* Pathology of Laboratory Rats and Mice. E. Cotchin and F. J. C. Roe, eds., Blackwell Scientific Publications, Oxford and Edinburgh.

Paterson, J. S. (1967). The Guinea-pig or Cavy. *In* The U.F.A.W. Handbook on the care and management of Laboratory Animals, pp. 241–287, E. and S. Livingstone Ltd., Edinburgh and London.

Porter, G. (1967). The Norway Rat. *In* The U.F.A.W. Handbook on the care and management of Laboratory Animals, pp. 353–390, E. and S. Livingstone Ltd., Edinburgh and London.

Radford, C. D. (1954). The larval genera and species of "Harvest mites" (Acarina: Trombiculidae). *Parasitology.* **44**, 247–276.

Richards, W. S. (1950). The distribution and biology of the harvest mite (Trombiculidae, Acarina). *Parisitology.* **40**, 118.

Rothschild, N. C. (1915). A synopsis of the British Siphonaptera. *Entomologist's mon. Mag.*, **51**, 49–112.

Smit, F. G. A. (1957a). *Handbook for the identification of British Insects: Siphonaptera.* Vol. I, part 16. Royal Entomological Society of London.

Smit, F. G. A. (1957b). The recorded distribution and hosts of Siphonaptera in Britain. *Entomologist's Gaz.* **8**, 45–75.

Sprent, J. F. A. (1963). *Parasitism.* University of Queensland Press, St. Lucia, Queensland, Australia.

Staats, J. (1966). The Laboratory Mouse, 1–9. *In* Biology of the laboratory mouse E. L. Green, ed., McGraw-Hill Book Company, New York and London.

Thomson, G. B. (1935). The parasites of British birds and mammals. IV. Records of mammal parasites. *Entomologist's mon. Mag.* **71**, 214–219.

Tuffery, A. A. and Innes, J. R. M. (1963). Diseases of Laboratory Mice and Rats. *In* Animals for Research, W. Lane-Petter, ed., Ch. 3. pp. 47–108, Academic Press, London and New York.

U.F.A.W. (1967). The U.F.A.W. Handbook on the care and management of Laboratory Animals, 3rd ed., E. and S. Livingstone Ltd., Edinburgh and London.

Van Duijn, C. (1956). Diseases of Fish. *Wat. Life*, London.

Wantland, W. W. (1955). Parasitic fauna of the Golden Hamster. *J. dent. Res.* **34**, 631–649.

Whitney, R. (1963). Hamsters. *In* Animals for Research, W. Lane-Petter, ed., Academic Press, London and New York.

Symp. zool. Soc. Lond. (1968) No. 24, 221–248.

THE WILD RUMINANT AS RESERVOIR HOST OF HELMINTH INFECTION

ANGUS M. DUNN

Department of Pathology, The University Veterinary School, Glasgow, Scotland

SYNOPSIS

Examination of the helminth parasite fauna of a number of wild ruminants indicates a high degree of cross-transmissibility to domesticated hosts. The extent of transmission in nature is moderated by a number of factors. Although few helminth parasites of ruminants are highly host-specific, the majority of those recorded from wild species have domesticated ruminants as their prime hosts. The effect of geographical location on the specific composition of infections is considered to be important in the general epidemiology of ruminant helminthiasis. Imperfect original description of some species and presumptive identification of others complicate the assessment of exchangeability of helminth populations between domesticated and wild animals. The possible emergence of strains and subspecies of differing infectivity hinders the acceptance of equal transmissibility of certain parasites to both groups of hosts. The majority of helminth infections of wild species to which domesticated stock are exposed are of low pathogenicity and the wild animal appears more likely to suffer in any encounter in communal grazing.

Of the helminth zoonoses attributable to ruminants, the important sources are usually in domesticated species; hydatidosis is the notable exception and some aspects of its sylvatic cycle are discussed.

It is concluded that, except in a few special cases, the background of helminthiasis in wild ruminants is not, at present, an important contributory factor in the epidemiology of helminthiasis in domestic stock.

INTRODUCTION

The free-living wild ruminants are so numerous and varied that they cannot be considered as a group any more than can our domesticated species. The zoological gap between the giraffe and its neighbour, the wildebeeste, or between the bison and the pronghorn, is far greater even than that between cattle and sheep. Not only are they taxonomically dispersed, but their different grazing behaviour and geographical locations bring the various species into greater or less contact with man and his animals. Communal grazing occurs in many regions of which the most notable are probably the game ranges of central Africa, where many gazelline species share grazing with cattle, sheep and goats; eastern Russia, where the dzeren and saiga antelope graze with sheep; the white-tailed deer areas of North America and the parts of Europe, of which the most extreme examples are probably in Scotland, where the deer are not confined to timbered forest but share grazing, and especially wintering, with the mountain sheep. Even in other areas, however, the privations of winter or extreme drought may drive the wild ruminant from its own ground on to the more sheltered or better irrigated pastures of domesticated stock.

The documentation of the endoparasitic fauna of most wild ruminants is still incomplete and it is out of the question to attempt a definitive statement on the epidemiological significance of their infections. Even in our present inadequate state of knowledge, however, the inevitable conflict between conservation of the wild species and growing pressure for grazing space for domesticated stock makes the significance of the wild carrier a question of some importance and one which cannot be evaded, even although no complete answer is yet possible.

Wild ruminants may act as reservoirs of helminth infection for domesticated stock and for man himself but, with increasing awareness that the wild animals have a positive economic status of their own, the risk which these animals present as reservoirs of infection for each other should also be acknowledged.

CROSS TRANSMISSABILITY OF HELMINTH PARASITES OF WILD AND DOMESTICATED RUMINANTS

The wild ruminants whose endoparasitic fauna are detailed below are selected not because they are completely representative but because they are the only ones which have considerable modern and critical documentation. It is fortuitous that they include some species which are numerous and which are in close contact with man and his animals. Information on other wild species is so incomplete that it is impossible to make even an approximate assessment of their significance and for many, including some of the African gazelles which have a manifest connection with the present subject, there are no records at all.

The information in the following lists is intended to indicate the extent of exchangeability of parasite populations and is compiled from many authors; it includes unpublished findings by the present writer.

Those helminth species marked with the letter d are also recorded from domesticated cattle, sheep or goats and those with D have domesticated ruminants as their prime hosts.

Family BOVIDAE

SUBFAMILY BOVINAE

Bison bison

The range of the American bison includes Canada and U.S.A. During the nineteenth century the population was reduced, largely by hunting, from over 60 million to 541 animals. The numbers have now recovered to more than 21 000.

Nematoda
Trichostrongylus axei
Ostertagia ostertagi
O. bisonis
Cooperia oncophora
C. mcmasteri
Grosspiculagia lyrata
Haemonchus contortus
Dictyocaulus viviparus
Oesophagostomum radiatum
Artionema labiato-papillosa
Trichuris ovis
Cestoda
Moniezia benedeni
Trematoda
Fasciola hepatica

Bison bonasus

The European bison, or wisent, is now extinct as a totally free wild animal. Apart from those in zoological collections there are small groups in reservations in eastern Europe. Poland, with five reserves and a few animals at supervised liberty in the forest of Bialowieza, has a stock of about 130 animals, representing about half the world population of European bison. Behaviourally, *B. bonasus* differs from *B. bison* in forming only small herds, usually of fewer than 30, and in undertaking no seasonal migration.

Nematoda
D *Trichostrongylus axei*
D *T. falculatus*
D *Ostertagia ostertagi*
D *O. circumcincta*
D *O. trifurcata*
D *O. orloffi*
D *O. bakuriani*
D *Grosspiculagia lyrata*
d *Spiculopteragia spiculoptera*
D *Cooperia oncophora*
D *C. punctata*
D *C. mcmasteri*
D *C. zurnabada*
D *Haemonchus concortus*
D *H. placei*
D *Nematodirus filicollis*
D *N. helvetianus*
D *N. spathiger*
D *Dictyocaulus filaria*
D *D. viviparus*
D *Oesophagostomum radiatum*
D *Oe. venulosum*

D *Chabertia ovina*
D *Bunostomum trigonocephalum*
D *B. phlebotomum*
D *Muellerius capillaris*
D *Strongyloides* spp.
D *Neoascaris vitulorum*
D *Gongylonema pulchrum*
D *Thelazia gulosa*
D *Th. skrjabini*
D *Th. rhodesii*
D *Artionema labiato-papillosa*
D *Trichuris ovis*
D *T. globulosa*
D *T. skrjabini*
D *Capillaria bovis*
D *C. bilobata*

Cestoda
D *Moniezia expansa*
D *M. benedeni*
D *Taenia hydatigena* larva
D *T. multiceps* larva
D *Echinococcus granulosus* larva

Trematoda

D *Fasciola hepatica*   D *Dicrocoelium dendriticum*
D *F. gigantica*   D *Paramphistomum cervi*

Subfamily Antilopinae

Gazella subgutturosa

The dzeren, or Persian, or goitred gazelle, is an Asiatic species inhabiting mostly desert and semi-desert and probably occurring in greatest numbers in Kazakhstan, Uzbekhistan and Azerbaijan.

Nematoda
D *Trichostrongylus capricola*
D *T. protolurus*
D *T. colubriformis*
D *T. vitrinus*
D *Ostertagia circumcincta*
D *Ostertagie circumcincta*
D *Grosspiculagia occidentalis*
D *G. trifida*
D *Marshallagia marshalli*
D *Camelostrongylus mentulatus*
D *Haemonchus longistipes*
D *Nematodirus spathiger*
D *N. abnormalis*
d *N. dogieli*
d *N. gazellae*
D *N. mauretanicus*
  N. mugosaricus
d *N. oiratianus*

D *Nematodirella cameli*
D *N. l. longispiculata*
D *Dictyocaulus filaria*
D *Chabertia ovina*
D *Skrjabinema ovis*
D *Gongylonema pulchrum*
D *Parabronema skrjabini*
D *Artionema labiato-papillosa*
  A. transcaucasica
  A. mugani
D *Trichuris ovis*
D *T. skrjabini*
Cestoda
D *Avitellina centripunctata*
D *Taenia hydatigena* larva
D *T. ovis* larva
Trematoda
D *Fasciola hepatica*
D *Dicrocoelium dendriticum*

Subfamily Caprinae

Tribe Saigini

Saiga tatarica

The saiga antelope in the wild state appears to be confined to Russia. It occurs in western and central Asiatic Russia in the salt semi-desert areas east of the Caspian Sea, principally in Kazakhstan, Uzbekhistan and Turkmenia. It is increasing in numbers and is cropped for food, upward of 20 000 being taken annually.

Nematoda
D *Trichostrongylus axei*
D *T. colubriformis*
D *Camelostrongylus mentulatus*
d *Nematodirus dogieli*

D *Dictyocaulus filaria*
  Skrjabinodera saiga
Cestoda
D *Moniezia expansa*
D *Thysaniezia giardi*

TRIBE RUPICAPRINI

Rupicapra rupicapra

The range of the chamois extends from the mountains of western Europe—the Pyrenees, Alps and Appenines—through the Carpathians to south eastern Europe and western Asia. It is a variable species and at least nine different races have been described from different parts of its range. Most of the records below are from Caucasian chamois.

Nematoda
D *Trichostrongylus colubriformis*
D *T. axei*
D *T. skrjabini*
D *Ostertagia circumcincta*
D *O. trifurcata*
d *O. leptospicularis*
  O. belockani
D *Grosspiculagia lyrata*
D *G. occindetalis*
D *Marshallagia marshalli*
  Rinadia mathevossiani
d *Spiculopteragia spiculoptera*
D *Teladorsagia davtiani*
D *Haemonchus contortus*
D *Nematodirus filicollis*
D *N. spathiger*
D *N. abnormalis*
d *N. davtiani*
  N. junctispicularis
d *N. oiratianus*

D *Dictyocaulus filaria*
D *D. viviparus*
D *Oesophagostomum venulosum*
D *Bunostomum trigonocephalum*
D *Chabertia ovina*
D *Muellerius capillaris*
D *M. tenuispiculatus*
D *Neostrongylus linearis*
D *Protostrongylus rufescens*
  P. rupicaprae
D *Spiculocaulus austriacus*
  Gelanocaulus boevi
D *Trichuris ovis*
D *T. globulosa*
D *T. skrjabini*
D *Capillaria bovis*
Cestoda
D *Moniezia expansa*
Trematoda
D *Fasciola hepatica*
D *Dicrocoelium dendriticum*

TRIBE CAPRINI

Ovis musimon

The range of the mouflon includes Corsica, Sardinia and parts of central Europe.

Nematoda
D *Trichostrongylus axei*
D *T. capricola*
D *T. colubriformis*
D *T. probolurus*
D *T. skrjabini*
D *T. vitrinus*
D *Ostertagia circumcincta*
d *O. davtiani*

D *O. dahurica*
D *O. trifurcata*
d *Grosspiculagia podjapolskyi*
D *Haemonchus contortus*
D *Nematodirus filicollis*
D *Dictyocaulus filaria*
D *D. viviparus*
D *Oesophagostomum venulosum*
D *Chabertia ovina*

M

D *Bunostomum trigonocephalum*
D *Muellerius capillaris*
D *Protostrongylus rufescens*
D *P. kochi*
D *Cystocaulus ocreatus*
   Capreocaulus capreoli

D *Gongylonema pulchrum*
D *Trichuris ovis*
D *T. globulosa*
D *T. skrjabini*
   Cestoda
D *Moniezia expansa*

Ovis ammon

The largest of sheep, the argali, inhabits semi-desert from the northern slopes of the Himalayas to the Kamchatka peninsula in far eastern Russia. It is most prevalent in Outer and Inner Mongolia, southern Siberia, Tibet and Sinkiang, which all border on the Gobi desert. Argali tend to herd in large groups and prefer lower ground. They appear to be indigenous to desert.

The typical form occurs in the Altai mountain area and there are two races, *O. a. polii*, or Marco Polo's sheep, in the Pamir and *O. a. hodgsoni* in Tibet and Kashmir.

Nematoda
D *Trichostrongylus axei*
D *Haemonchus contortus*
   Nematodirus archari
d *N. dogieli*
D *Dictyocaulus filaria*
   Varestrongylus pneumonicus
D *Protostrongylus raillieti*

D *P. hobmaieri*
D *P. davtiani*
D *P. skrjabini*
D *Spiculocaulus leuckarti*
D *Cystocaulus nigrescens*
   Cestoda
D *Moniezia benedeni*
D *Taenia hydatigena* larva

Ovis canadensis

The bighorn sheep of North America is probably of Asiatic origin. Its distribution includes, as well as North America, north eastern Asia, and it is thought to have crossed to America by the Bering isthmus. It has various races, the typical being the brown form in British Columbia, and the range of colour extends to the white form of the Yukon.

Nematoda
D *Trichostrongylus axei*
D *Grosspiculagia occidentalis*
D *Pseudostertagia bullosa*
D *Marshallagia marshalli*
D *Cooperia oncophora*
D *Haemonchus placei*
D *Nematodirus abnormalis*
   N. odocoilei
D *N. spathiger*

Protostrongylus rushi
P. stilesi
D *Skrjabinema ovis*
   Trichuris sp.
   Cestoda
D *Moniezia benedeni*
   Wyominia tetoni
D *Taenia hydatigena* larva

FAMILY CERVIDAE

SUBFAMILY MOSCHINAE

Moschus moschiferus

The musk deer is the most primitive of all deer and is of solitary habit. Its range extends from almost 72° North, within the Arctic Circle, south to Korea, Assam and Burma, and includes Siberia, Mongolia, Tibet and western China. It is found at altitudes of up to 12 000 feet in the mountains of Central Asia. This little deer has some economic importance, the male secreting musk which is used as a basis for perfumes and is collected by cannulation of the gland. In some regions it is still hunted for the musk and has been exterminated from many parts of its range.

Nematoda	d *Artionema kabargi*
D *Trichostrongylus axei*	D *Trichuris ovis*
D *T. colubriformis*	D *Capillaria bovis*
D *Ostertagia ostertagi*	Cestoda
D *Nematodirus filicollis*	D *Moniezia expansa*
Pneumocaulus kadenazii	D *Taenia hydatigena* larva
d *Pygarginema skrjabini*	

SUBFAMILY CERVINAE

Dama dama

Although its primaeval territory was the forest of the Mediterranean Basin, the fallow deer has been introduced to most parts of Europe. Indigenous fallow still occur on the southern coast of Asia Minor and, very rarely, in the African part of their original range.

Nematoda	D *Skrjabinagia kolchida*
D *Trichostrongylus axei*	*Apteragia quadrispiculata*
D *T. colubriformis*	D *Cooperia curticei*
D *T. vitrinus*	D *C. oncophora*
D *Ostertagia circumcincta*	D *C. pectinata*
d *O. leptospicularis*	D *C. punctata*
O. mossi	D *Haemonchus contortus*
D *O. trifurcata*	D *Nematodirus filicollis*
d *Grosspiculagia lasensis*	D *N. helvetianus*
D *Teladorsagia davtiani*	*N. roscidus*
Spiculopteragia asymmetrica	D *N. spathiger*
d *S. spiculoptera*	D *Dictyocaulus filaria*
Rinadia mathevossiani	D *D. viviparus*

Oesophagostomum sikae
D Oe. venulosum
D Chabertia ovina
D Muellerius capillaris
  Bicaulus sagittatus
D Gongylonema pulchrum
  Wehrdikmansia flexuosa
  W. cervipedis
  Trichuris ovis

D Capillaria bovis
  Cestoda
D Moniezia expansa
D Taenia hydatigena larva
  Trematoda
D Fasciola hepatica
D Dicrocoelium dendriticum
D Paramphistomum cervi

Cervus elaphus

The range of the red deer in its various forms includes much of Europe, in which it is the most important big game animal, and Asia.

Nematoda
  Trichostrongylus askivali
D T. axei
D T. capricola
D T. colubriformis
D Ostertagia circumcincta
d O. leptospicularis
  O. mossi
D O. ostertagi
d Grosspiculagia lasensis
D G. lyrata
  Spiculopteragia asymmetrica
d S. spiculoptera
  Rinadia caucasia
  R. mathevossiani
D Skrjabinagia kolchida
  Apteragia quadrispiculata
D Cooperia curticei
D C. pectinata
D Nematodirus filicollis
  N. roscidus
D Dictyocaulus filaria
D D. viviparus
  D. eckerti

  Oesophagostomum cervi
D Oe. radiatum
  Oe. sikae
D Oe. venulosum
D Chabertia ovina
D Bunostomum trigonocephalum
D B. phlebotomum
  Bicaulus sagittatus
  Elaphostrongylus cervi
  Wehrdikmansia flexuosa
  W. cervipedis
  Parafilaria antipini
  Artionema altaica
  Cestoda
D Moniezia expansa
D M. benedeni
D Taenia hydatigena larva
D T. multiceps larva
D Echinococcus granulosus larva
  Trematoda
D Fasciola hepatica
D Dicrocoelium dendriticum
D Paramphistomum cervi

Cervus nippon

The sika has several races including the Manchurian, Japanese-southern Chinese and the Taiwan. It has been introduced into many parts of Europe and is probably the most amenable of all deer to colonization in different climates and dietary conditions. Introductions almost invariably settle well and thrive and, in northern Europe, form hybrid crosses with red deer. Its indigenous range is limited to deciduous forest in the southern part of eastern Asia. Like other small deer, snow depth limits its extension and it is not found in regions where the mean depth exceeds 10 to 20 cm.

The antlers are highly valued, even more than those of the Asiatic wapiti, or maral, as a so-called "tonic elixir" and, like the latter animal, the sika is farmed for these in China and Russia. The antlers are sawn off while still young and filled with blood.

Nematoda
D *Trichostrongylus axei*
D *T. colubriformis*
D *Ostertagia circumcincta*
D *O. gruhneri*
d *O. leptospicularis*
D *O. ostertagi*
d *Grosspiculagia lasensis*
D *G. lyrata*
  Spiculopteragia asymmetrica
d *S. spiculoptera*
  Rinadia mathevossiani
D *Skrjabinagia kolchida*
D *Cooperia pectinata*
D *Haemonchus contortus*
d *Nematodirus oiratianus*
D *N. spathiger*
  Ashworthius sidemi
D *Dictyocaulus filaria*
D *D. viviparus*
D *Oesophagostomum asperum*
D *Oe. radiatum*
  Oe. sikae

D *Oe. venulosum*
D *Globocephalus longemucronatus*
D *G. samoensis*
  Schulzinema miroljubovi
D *Bunostomum trigonocephalum*
  Elaphostrongylus panticola
  Pygarginema cervi
  Wehrdikmansia cervipedis
  W. flexuosa
  Artionema altaica
D *Trichuris ovis*
D *T. skrjabini*
D *Capillaria bovis*
Cestoda
D *Moniezia benedeni*
D *Taenia hydatigena* larva
D *Echinococcus granulosus* larva
Trematoda
D *Fasciola hepatica*
D *Dicrocoelium dendriticum*
D *Eurytrema pancreaticum*
D *Paramphistomum cervi*
  Cotylophoron sp.

SUBFAMILY ODOCOILEINAE

TRIBE ODOCOILEINI

Odocoileus virginianus

The white-tailed deer ranges from Canada to central and northern South America, with a decrease in size toward the tropical part of the range.

Nematoda
D *Trichostrongylus colubriformis*

D *Ostertagia circumcincta*
  O. mossi

O. odocoilei
D Grosspiculagia lyrata
D Cooperia pectinata
D C. punctata
D Haemonchus contortus
D H. similis
D Nematodirus filicollis
D N. spathiger
D Dictyocaulus filaria
D D. viviparus
D Oesophagostomum venulosum
D Chabertia ovina
  Eucyathostomum longesubulatum
  Protostrongylus coburni
d Elaphostrongylus tenuis
  Leptostrongylus alpenae
D Skrjabinema ovis

D Gongylonema pulchrum
D G. verrucosum
  Artionema tundra
  Wehrdikmansia cervipedis
D Trichuris ovis
            Cestoda
D Moniezia expansa
D M. benedeni
D Thysanosoma actinoides
D Taenia hydatigena larva
D Echinococcus granulosus larva
            Trematoda
d Fascioloides magna
D Dicrocoelium dendriticum
D Paramphistomum cervi
  P. liorchis

TRIBE ALCINI

Alces alces is the largest living cervid. It is an animal of forests and marshes and inhabits almost the entire taiga region of Europe, Asia and North America. It is rarely found far from water. Because of its long legs it can survive in deep snow but its somewhat giraffe-like conformation hinders it from grazing and it does so only when compelled to by privation. It is essentially a browser, supplementing its leafy diet with bark. It is not a highly gregarious animal, groups usually consisting of five or ten animals, and it is often solitary or in pairs.

Alces a. alces occurs in western Siberia, in European Russia, in Scandinavia and, sparsely, in Poland. Once in some danger, the numbers of the European elk are increasing as a result of protection.

Alces a. americana. the American moose, is found in the taiga of North America and the northern far east of Russia.

Alces a. alces

            Nematoda
D Trichostrongylus axei
D T. capricola
D T. colubriformis
D T. vitrinus
  Ostertagia antipini
d O. leptospicularis
D O. orloffi
  Spiculopteragia alcis
d S. dagestanica

d S. spiculoptera
  Rinadia mathevossiani
  Skrjabinagia lyrataeformis
D Haemonchus contortus
  Nematodirella alcidis
D N. cameli
D N. longissimespiculata
D Dictyocaulus viviparus
D Bunostomum trigonocephalum
  Bicaulus alces

D *Oxyuris equi*
D *Trichuris ovis*
D *Capillaria longipes*
Cestoda
D *Moniezia benedeni*
D *Thysanosoma actinoides*
Taenia cervi larva

D *T. hydatigena* larva
D *T. krabbei* larva
D *Echinococcus granulosus* larva
Trematoda
d *Parafasciolopsis fasciolaemorpha*
D *Paramphistomum cervi*

Alces a. americana

Nematoda
D *Trichostrongylus axei*
D *Haemonchus contortus*
Nematodirella alcidis
D *N. longispiculata*
Dictyocaulus eckerti
D *D. viviparus*
d *Elaphostrongylus tenuis*
Wehrdikmansia cervipedis

Cestoda
D *Moniezia benedeni*
D *Thysanosoma actinoides*
D *Taenia hydatigena* larva
D *T. krabbei* larva
D *Echinococcus granulosus* larva
Trematoda
d *Fascioloides magna*
D *Paramphistomum cervi*

TRIBE CAPREOLINI

Capreolus capreolus

The roe deer ranges over most of Europe and Asia and several strains or subspecies are recognized. It is usually a forest animal, browsing for the most part and grazing only intermittently although it can survive well on treeless range.

Its northern limit was once supposed to be the tundra but it is now accepted that depth of snow is the critical factor, its general northern distribution and its vertical distribution in the mountains of Asia being confined within an average maximum depth of 50 cm.

Nematoda
D *Trichostrongylus axei*
T. askivali
D *T. capricola*
D *T. colubriformis*
D *T. longispicularis*
T. retortaeformis
D *T. skrjabini*
Ostertagia antipini
D *O. circumcincta*
d *O. davtiani*
d *O. leptospicularis*
D *O. ostertagi*

d *Grosspiculagia lasensis*
D *G. lyrata*
D *G. occidentalis*
Spiculopteragia alcis
S. asymmetrica
D *S. dagestanica*
d *S. spiculoptera*
Rinadia mathevossiani
D *Skrjabinagia kolchida*
S. lyrataeformis
Capreolagia skrjabini
D *Teladorsagia davtiani*
Apteragia quadrispiculata

D *Marshallagia marshalli*
D *Cooperia pectinata*
D *C. punctata*
D *Haemonchus contortus*
D *H. placei*
D *Nematodirus battus*
D *N. filicollis*
D *N. helvatianus*
   N. junctispicularis
   N. roscidus
   Dictyocaulus eckerti
D *D. filaria*
D *D. viviparus*
   Oesophagostomum cervi
   Oe. sikae
D *Oe. venulosum*
D *Chabertia ovina*
D *Bunostomum phlebotomum*
D *B. trigonocephalum*
D *Protostrongylus rufescens*
D *Muellerius capillaris*
D *Spiculocaulus austriacus*
   Capreocaulus capreoli
D *Ascaris lumbricoides*
   Pygarginema skrjabini
D *Gongylonema pulchrum*

   Artionema capreoli
d *A. kabargi*
D *A. labiato-papillosa*
   Wehrdikmansia rugosicaunda
d *Trichuris capreoli*
D *T. globulosa*
D *T. ovis*
D *T. skrjabini*
D *Capillaria bovis*
   Cestoda
D *Moniezia benedeni*
D *M. expansa*
   Avitellina pygargi
   Avitellina arctica
D *Thysanosoma actinoides*
   Taenia cervi larva
D *T. hydatigena* larva
D *T. krabbei* larva
D *T. multiceps* larva
D *Echinococcus granulosus* larva
   Trematoda
D *Fasciola hepatica*
d *Parafasciolopsis fasciolaemorpha*
D *Dicrocoelium dendriticum*
D *Paramphistomum cervi*

Family ANTILOCAPRIDAE

Antilocapra americana

The pronghorn is the sole representative of its family and, although often so termed, is not a true antelope, these being confined to the Old World. It tends to inhabit rocky, semi-desert country and is found in south western U.S.A.

   Nematoda
D *Trichostrongylus axei*
D *T. capricola*
D *T. colubriformis*
d *Ostertagia bisonis*
D *O. circumcincta*
D *O. ostertagi*
D *Grosspiculagia occidentalis*
D *Pseudostertagia bullosa*
D *Marshallagia marshalli*

D *Cooperia oncophora*
D *C. pectinata*
D *C. punctata*
D *Haemonchus contortus*
D *Nematodirus abnormalis*
D *N. filicollis*
   N. lanceolatus
D *N. spathiger*
D *Nematodirella longispiculata*
   Protostrongylus macrotis

D *Trichuris discolor*

D *Capillaria brevipes*

D *C. bovis*

Cestoda

D *Moniezia benedeni*

D *M. expansa*

D *Thysanosoma actinoides*

An obvious omission from these lists is the caribou. It is not presented here although, it is well documented, because total transmissibility of helminth parasites with its domesticated form, the reindeer, may be presumed.

One difficulty in approaching the available data on the specific composition of the helminth infections of wild ruminants is that many records are inadequate. Older parasites have disappeared into synonymy as, for example, *Trichostrongylus andreevi*, *Ostertagia hamata*, *O. pinnata*, *Haemonchus okapiae*, *Spiculopteragia bohmi* and many *Moniezia* species. Others have suffered from imperfect original description and cannot now be recognized as, for instance, *Ostertagia cervi* (Cameron, 1931) from the Scottish red deer which has been variously allocated to the genera *Skrjabinagia* and *Spiculopteragia* by subsequent writers. It has never been found by the present writer in the examination of more than 1000 Scottish red deer and, although not, perhaps, relevant to the present discussion, he considers that this parasite, described by Cameron as "the commonest parasite of the abomasum of red deer in Scotland", was, in all probability, *Spiculopteragia spiculoptera*, which is, indeed, very common and resembles superficially Cameron's parasite. It is something of an irony that *S. spiculoptera*, as a new species, did not appear in published description until three years after Cameron's doubtful *O. cervi* (Gushanskaja, 1934).

FACTORS MODIFYING THE ACQUISITION OF PARASITES

Specificity

In general the parasites of ruminants are not notably host-specific and the possibility of restriction to a single species or even genus is slight and new records are diminishing even the present small numbers of wholly specific parasites. It is among the protostrongylids that the narrowest ranges are found—*Protostrongylus coburni* only in the northern white-tail, *Leptostrongylus alpenae* only in the white-tail, *Pneumocaulus kadenazii* only in musk deer, *Bicaulus alces* only in the European elk and *Capreocaulus capreoli* only in the roe.

Restriction of the host range at subfamily or family level, although more common, still includes only a minority of the ruminant helminths. Parasites in the foregoing lists which are confined to cervids are the trichostrongyloids, *Trichostrongylus askivali*, *Ostertagia antipini*, *O.*

mossi, O. odocoilei and *Rinadia mathevossiani;* the strongyloids, *Eucya-thostomum longesubulatum, Oesophagostomum sikae* and *Oe. cervi;* the metastrongyloids, *Bicaulus sagittatus* and *Elaphostrongylus panticola;* the filarioids, *Artionema tundra* and *Wehrdikmansia cervipedis;* the cestodes, *Avitellina arctica, A. tundra* and *Taenia krabbei* larvae; and the single trematode, *Paramphistomum liorchis.* Those confined to caprines are even fewer and are all protostrongylid—*P. hobmaieri, P. raillietti, P. skrjabini, P. kochi, Spiculocaulus leuckarti, Muellerius capillaris* (the only report from a cervid was based on larval identification), *Cystocaulus nigrescens* and *Varestrongylus pneumonicus.*

Helminths with almost unrestricted ranges throughout ruminants are all well known and include, for example, most trichostrongyloids, *Gongylonema pulchrum, Artionema labiato-papillosa, Thelazia* spp., *Trichuris ovis, Capillaria bovis, Moniezia* spp., unilocular hydatid, *Cysticercus tenuicollis, Fasciola hepatica* and *Paramphistomum cervi.* The reservation must be made, however, that some of these apparently wide ranges of familiar parasites may be due to doubtful attribution, as discussed above.

The records in these lists of *Ascaris lumbricoides* from *Capreolus capreolus, Globocephalus longemucronatus* from *Cervus nippon* and the highly host-specific *Oxyuris equi* from *Alces a. alces* must be treated with reserve.

Dictyocaulus presents, perhaps, the best illustration of the problems which beset the modern worker. Not only has it been subject in the past to improper recognition, which accounts for the creation and later collapse of the two species *D. noerneri* and *D. hadweni,* but there is no doubt that, as in many other cases, much of the evidence on the epidemiology of its species has been based on presumptive identification. The presence of a large nematode high in the bronchi and trachea of a ruminant indicates, almost beyond doubt, the generic identity of *Dictyocaulus.* What has bedevilled the subject has been the extension of this generic identification to species, purely on a host basis. Hence, depending upon the whim of the investigator, the parasites have been allocated to the species common in the host with which he, perhaps, happened to be familiar, *D. filaria* or *D. viviparus.* Challenge to these casual allocations came with the proposed cervine species, *D. eckerti* (Skrjabin, 1931) which, originally described from the reindeer, has an extensive distribution in the subfamily (caribou, maral, wapiti, moose, red and roe deer). The genus has been divided into the subgenera *Dictyocaulus* and *Micrurocaulus,* based on the fusion (*D. (D.) filaria* and *D. (D.) arnfieldi*) or non-fusion (*D. (M.) viviparus, D. (M.) cameli* and *D. (M.) eckerti*) of the postero-lateral and medio-lateral bursal rays along their whole length. Thus, *D. viviparus* and *D. eckerti* are in the same subgenus. It must be conceded that the morphological evidence for the specific discreteness of the two latter species is slight and it has so far been accepted mainly on a biological basis although the writer has encountered apparent cases of bovine-cervine transmission in the field, the cervines suffering in the encounter.

Whatever diffidence there may be in accepting *D. eckerti*, however, there is no question that the use of "*D. filaria*" and "*D. viviparus*" as "catch-all" species must be forgotten. It should be added that in the "catch-all" category may be included "*Trichuris ovis*" and the ubiquitous "*Paramphistomum cervi*".

The situation exists where certain parasites, and in particular the trichostrongyloids, although including both domesticated and wild ruminants in their ranges, require the continuous presence of a reservoir of infection to achieve this.

The reservoir may be in wild animals, infection of domesticated stock resulting from casual encounter but hardly ever amounting to important proportions. Thus, *Ostertagia leptospicularis* and *Grosspiculagia lasensis*, essentially cervid parasites, have been found by the writer in feral goats grazing the terrain of deer and even as a majority infection in sheep by Jansen (1963). The same author (1960) indicates that the very common *Trichostrongylus axei* may be in this category, having noted differences in the populations in roe from different areas, those from regions where domesticated stock are sparse being unable to develop to large numbers in the deer. He also suggests that *O. ostertagi* and *G. lyrata* have the same epidemiological character; the writer would agree with this and would add *O. circumcincta*, which he has found to be common in red deer from sheep areas but rare or absent in animals from sheep-free areas. Whitlock (1939) noted that although *Dictyocaulus* spp. and *Protostrongylus rufescens* both occurred in white-tailed deer in Michigan, they were only found in animals which shared grazing with sheep. It is possible that many other metastrongyloids may have the same necessity for a continuous background of infection in a prime host for their perpetuation in other species. Boev (1957), for example, has allocated *P. hobmaieri*, *P. raillieti*, *P. skrjabini*, *Bicaulus schulzi* and *Cystocaulus nigrescens* as specific parasites of sheep which may facultatively parasitise other animals. Similarly, *Spiculopteragia* spp., are primarily cervid parasites but have been found occasionally in domesticated stock and even in the closely preserved European bison (Drozdz, 1961). *Pseudostertagia bullosa* appears to have the pronghorn as its prime host (Lucker and Dikmans, 1945), infection of domesticated and bighorn sheep being acquired from this reservoir; it has not, however, been recorded from domesticated cattle which share the same range as the pronghorn. Although Jansen (1960) suggests that *Apteragia quadrispiculata* infects red deer only when they share grazing with fallow, the writer cannot accept this, having found the parasite in red deer when they were the sole ruminants on an island forest. *Trichostrongylus retortaeformis* is primarily parasitic in rabbits and hares; it has been found in wild roe (Jansen, 1958; Dunn, 1965a) and red deer (A. M. Dunn, unpublished), and in captive Chinese water deer (*Hydropotes inermis*) (Leiper and Clapham, 1938) but may be presumed to be a casual parasite in these cervids.

In a different category, perhaps, are those parasites which can infect other species but which are restricted by the reaction of the host to their presence. *Elaphostrongylus tenuis*, comfortably tolerated by the white-tail, can be lethal for moose and sheep; *Fascioloides magna*, tolerated, again, by the white-tail, is isolated by the host reaction in cattle and is often fatal for sheep although it has been suggested (Campbell and Todd, 1954) that the surviving sheep may play a similar rôle to cervids in the dissemination of this parasite.

It may be conjectured that, at this moment in time, we are witnessing parasites in different rates of evolution. The oxyuroids, for example, appear to have slowed their evolutionary pace and are relatively stable, whilst many strongylates and spirurates appear to be in a process of rapid differentiation. Das and Whitlock (1960) suggest that *Haemonchus contortus* is in such a condition of rapid differentiation and, certainly, this species had already been separated into two morphological species (Roberts *et al.*, 1954), the parent and *H. placei*, after the different behaviour of strains from sheep and cattle had been noted. The differences in pathogenicity of *H. contortus* from different ruminant hosts demonstrated by Samson *et al.* (1964) may be an illustration of emerging subspecies. In *Dictyocaulus* it is not only possible that *D. eckerti* and *D. viviparus* are recently separated strains or species but that *D. filaria* is also in process of differentiation, Boev (1957) having indicated that in communal grazing of sheep and goats on infected pasture goats habitually harbour many fewer parasites than do sheep. It has been suggested, too, that among the metastrongyloids the same process may be happening; *Protostrongylus kochi*, for example, whose range of final hosts is in small caprines and cervines, appears to have adapted itself to rabbits and hares, forming a new biological, but not morphological, species, *P. cuniculorum* (Joyeux and Gaud, 1946) which cannot now infect ruminants. The writer would suggest that *Trichostrongylus colubriformis* and *T. retortaeformis*, the former in ruminants and the latter in rabbits and hares, which show some morphological similarity, may be recently separated species, which are now adding morphological distinction to their biological differences.

The importance of such differentiation would obviously be great in an epidemiological context but insufficient is known of its extent to allow its application to this discussion.

Geography

Many cases of specific resistance are only apparent, the parasite being absent from an animal simply because an encounter never takes place. Many cases of presumed specific resistance have broken down, for example, when animals have been brought to the unnatural conditions and broad infective substrates of zoological gardens. Thus, reindeer, in their northern habitats do not harbour protostrongylids because of the sparseness of the molluscan intermediate hosts in these areas, yet when they are brought to zoos they often show high levels of infection. *Rinadia* and *Spiculopteragia* are both fairly specific for cervids but Jansen (1963) has found them in a mouflon in a zoo. Similarly, the saiga antelope, at liberty in the semi-desert, has no lung nematode parasites but in zoos is often found infected with *Dictyocaulus* spp. (Boev, 1957). The same author points out that, even under natural conditions, geography plays an important part in the acquisition of parasites, the argali, for example, being much more heavily parasitized by protostrongylids in the mountain areas of Kazakhstan than in the semi-desert. In the same area, too, there is a contrast between the protostrongylid fauna of the Asiatic ibex which, living in the Syugaty mountains, has a high protostrongylid incidence and harbours different species, and the dzeren which, living in the adjacent valley, has none, the terrain in the valley, with low humidity and little vegetation being an unsuitable habitat for land molluscs, whereas the mountains, with high humidity and heavy vegetation provide an excellent habitat.

In the Betpak-Dala desert of Kazakhstan the dominant wild ruminant is the argali and it has been shown in that area that imported sheep, which show an initial drop in their general helminth infections, soon acquire a number of other species from the local wild ruminants. The hazard is well recognized by the Kazakhstanis, who are reclaiming the land for sheep grazing (Karabaev, 1953). Similarly, among wild ruminants themselves, Drozdz (1963) found that sika, introduced to Poland from the East, lost their own infections but acquired new populations from the native fallow deer.

Mönnig (1931), observing the communal grazing of antelope species with sheep on many farms in South Africa, carried out a series of experimental infections in which he transmitted *Trichostrongylus* spp.,

Cooperia spp., *Haemonchus* spp., *Nematodirus* spp. and *Strongyloides* spp. to sheep from blesbok (*Damaliscus dorcas*), waterbuck (*Kobus ellipsiprymnus*) and springbok (*Antidorcas marsupialis*).

In Uzbekhistan the dzeren has no specific parasites and its total population, excepting *Nematodirella cameli*, is exchangeable with the Karakul sheep of the area, so that it presents a continuous reservoir for the communally grazing sheep (Karmanova and Murtzaev, 1953).

On the other hand, an example of a parasite which is presented to a variety of hosts but confines itself to one group is *Avitellina arctica*. This cestode occurs exclusively in cervids although the ranges of its type host, the Siberian roe, *Capreolus c. pygargi*, and of its other principal hosts, *Rangifer t. uralensis* and *R. a. arcticus*, overlap with that of domesticated sheep.

Geographical location appears to play as important a part in the acquisition of alimentary parasites as any other factor, as indicated by the specific composition in the pronghorn, zoologically isolated but communally grazing with other hosts and sharing their parasites, and in the American bison and the wisent, where the parasite fauna in each animal is demonstrably geographically rather than zoologically characterized. The bighorn, however, shows zoological rather than geographical characters, for although this animal appears not to mix with domesticated sheep in its natural habitat, it has a parasite population totally exchangeable with the domesticated species.

Dietary habits would influence the aquisition of parasites with direct life cycles at least and it would be supposed that burdens would be slighter in animals whose preferences are for browsing rather than grazing. Most cervids, in fact, are browsers for most of the year and it is only when young grass is present, or when driven by the absence of browse in winter, that they turn to grazing. Among the family the browsing habit would appear to be almost obligatory in *Alces* spp. which are anatomically unsuited to grazing with their long legs and short necks and which subsist for the most part on shrubs, trees, bark and the longer grassy plants.

PATHOGENICITY OF INFECTIONS

Many of the earlier parasitological investigations were carried out with the identity of the helminths as the sole object and without any interest in the burdens present. Hence it is difficult to arrive at any assessment of the pathogenicity of many of the parasites and, in particular of the trichostrongyloids, whose simple presence in an animal means very little. Nevertheless, these old records have substantial value as true observations.

A much more serious impediment is the casual allocation of blame to helminths simply because they happened to be present in an animal at death or along with morbid change. Many parasites have been incriminated as pathogens on this unlikely basis. *Chabertia ovina*, for example, which is rarely a pathogen except in intensively husbanded sheep, has been named as the cause of death of roe in Switzerland (Grieder, 1934) and in the former East Prussia (Hübner, 1938) and as a cause of antler malformation in deer (de Nahlik, 1959). It is doubtful whether much dependence should be placed on such reports of lungworm infection as those of Krems (1939) where parasites, usually *Protostrongylus capreoli* but occasionally *Dictyocaulus viviparus*, were found in 101 Bavarian roe and are stated to have "probably" killed 34 of these when *Protostrongylus* is not recognized as a notably pathogenic genus. Nor is it easy to accept his opinion that of 51 deer in which *Haemonchus* was found, 37 were "probably" killed by the worm; while it is a substantially pathogenic genus, such mortality is not often seen even in heavily crowded domesticated sheep. Hence his conclusion that parasites "probably" killed 30% of the deer he examined must be treated with some reserve, as must others of the same type.

It is especially true of helminthiasis that a careful distinction must be made between the presence of infection and the presence of disease due to that infection. Most helminth infections, even in closely observed domesticated stock, produce no detectable clinical or pathological change and, of course, it is usually impossible to study sublethal parasitic infections in the wild ruminant so that the margin between infection and apparent disease is much wider. The information summarized below, it will be agreed, demonstrates at least the extent of the study still required on the pathogenicity of helminthiasis in the wild ruminant.

Trichostrongylidosis can be a serious problem in domesticated stock and there is no reason to suppose that wild ruminants with comparable burdens would be any less severely affected. It is a fact, however (Dunn, 1965a, 1965b), that the host-trichostrongylid association appears to be a well adjusted one in wild ruminants and heavy burdens are not often encountered. Attribution of serious pathogenicity have mostly been made in North America. Heavy deer losses from *Ostertagia* spp., *Haemonchus* spp. and *Trichostrongylus* spp. have been noted in deer on the ranges of the northern Californian coast (Longhurst *et al.*, 1952), and severe gastritis due to *T. axei* has been encountered in the wapiti (Eveleth and Bolin, 1955). Although *Haemonchus contortus* is often blamed as a cause of mortality in mid- and southern European deer, perhaps, as suggested above, on occasion with limited justification, it has never been encountered as a primary pathogen, or even in substantial numbers, in the northern European countries. It is very rare in British deer and the writer has examined a number, grazing the same territories as heavily infected goats, which carried none of this species. The experiments of Samson *et al.* (1964), who showed that strains of this parasite from Barbary sheep and bighorn passaged into domesticated sheep were of considerably lower pathogenicity than the homologous strain, may indicate, as has already been mentioned in connection with specificity, that the parasite has less significance as a cross-transmissible species than has been thought.

In the writer's own examination of parasites of Canadian bison he has found evidence of moderate burdens of *O. ostertagi* and, whilst not themselves of primary pathogenic importance, it is possible that they might have an additive effect in trichostrongylidosis of young domesticated stock on bison range.

It has already been remarked that a number of parasites may require the continuous presence of a reservoir in domesticated animals to achieve more than rare status in the wild and in trichostrongylidosis some extension of this character into terms of actual pathogenicity may be possible, as shown by the epidemiological observations of Longhurst and Douglas (1953). These authors noted that deviations in feeding behaviour could contribute to the acquisition of pathogenic trichostrongylid burdens by Californian deer. Although by habit browsing animals, when the autumn rains came in November the deer moved on to the new growth of grass, which they shared with sheep, and grazed until mid-March when the new shoots appeared on the browse plants. The greatest losses in deer occurred during the November–March period and, once they had moved to browse which is almost inaccessible to helminth infective stages, losses decreased. Sheep on the same grassland wintered well, one group of 95 coming through without casualty although twelve deer died. It appears from these observations at least that the wild ruminants were the sufferers in communal grazing, acquiring predominantly ovine helminth species from the reservoir in the domesticated ruminants.

Dictyocaulus spp. are common in cervids and can cause fatal disease but the relevance of these infections as reservoirs for domesticated stock is in doubt until the cross-transmissibility of the caprine, bovine and cervine species is established.

Pulmonary metastrongyloidosis is common in domesticated sheep and goats as well as in wild sheep, goats and deer but in none of these animals is there any widespread evidence of important pathogenic effects although occasional outbreaks of verminous pneumonia in bighorn in U.S.A. are attributable to protostrongylids.

The cerebrospinal metastrongyloids are of considerable local importance and it is due to the findings of Anderson (1963a, 1964) and Anderson and Strelive (1967) that the significance of *Elaphostrongylus tenuis* as an accidental pathogen among wild ruminants themselves has been established. The worm is well tolerated by its prime host, the white-tailed deer, in which the adults live in the subdura following larval migration from abomasum. It appears that, in the prime host, not only are the adults tolerated in their predilection site, but the host is able to present a rapid defence to invading larvae so that few ever reach the central nervous system. In other cervids, however, and notably in the moose and wapiti, very large numbers of larvae reach nervous tissue and severe damage occurs. In the moose, motor ataxia, temperamental changes to fearlessness, lethargy and apparent blindness (or, at least, impaired vision) have been noted. In infection of sheep with this species the larvae do not appear to settle down subdurally but continue their wanderings in the C.N.S., causing encephalomalacia and local haemorrhage, a pathogenesis resembling that of ovine artionemiasis in the Old World. There are several reports of high incidences of inapparent infection in white-tail. Alibasoglu *et al.* (1961) found 75% of the 81 heads they examined in Pennsylvania infected; De Guisti (1963) found that, of 836 heads examined in Michigan, 56·46% were infected; Anderson (1963b) examined deer throughout the year in Algonquin Park and found an overall incidence of 41% with some seasonal variation (46% in summer and 24% in winter).

Of the other elaphostrongyloses, *E. rangiferi* occurs in the submeningeal region in reindeer in Scandinavia, causing paresis and convulsions (Ronéus and Nordkvist, 1962) and *E. panticola* causing occasional heavy losses in maral in Eastern Russia.

None of the spirurates which may originate from wild ruminants is of great pathogenic importance. *Artionema labiato-papillosa* is widespread in both domesticated and wild ruminants throughout the world but infection is usually discovered only at routine post-mortem examination; the cerebrospinal nematodiasis associated with paralysis in sheep and caused by the erratic wanderings of this parasite has its reservoir in domesticated ruminants.

Thelazia, like *Artionema*, is ubiquitous in both domesticated and wild animals and occurs in a wide range of hosts so that there can be no question of incriminating wild species as important reservoirs. *Elaeophora schneideri* has been found as an inapparent infection in mule deer in several western states of the U.S.A. but the prime host of the parasite is the domesticated sheep and deer cannot be considered a useful reservoir. *Wehrdikmansia cervipedis* is common in cervids in

America and has been noted in sika in Europe but has not so far been recorded from domesticated species. Ritcey and Edwards (1958) found that, in moose, heavy infections would extend from the usual site on the feet and brisket to the shoulders and sides. Although in one case three worms per square inch were noted over the brisket, no pathogenic effect was found. Although this parasite has had attributed to it by some authors pain, swelling, ulceration and loss of digits, others of extensive field experience, such as Herman (1947), have never found important lesions associated with it so that this may be another instance of a parasite being incriminated because it happened to be present fortuitously with bacterial infection of the region.

Fascioloides magna is one of the few parasites of wild animals which may be definitely incriminated as a pathogen for domesticated ruminants. Well tolerated by its prime cervid host even in the presence of considerable loss of liver tissue [Olsen (1949) found up to 31 of these flukes in single deer livers with the loss of a third of the hepatic tissue in apparently healthy animals], only a few parasites are necessary to cause severe and sometimes fatal damage in sheep from their unrestricted migration. Although there can be no question that almost the entire reservoir of infection for sheep is in the cervid prime hosts, Campbell and Todd, as has already been noted, have suggested that sheep surviving the infection may contribute to the spread of the parasite.

Fasciola hepatica is so widespread in ruminants and other animals that it is difficult to assess the significance of local reservoirs in wild ruminants. It has long been recognized that in many areas rabbits and hares may present a continuous background of endemicity and there is no doubt that in communal grazing a cervid reservoir remains a serious obstacle to control. In Africa the same problem may arise in the control of *F. gigantica*. Jansen (1963) suggests, however, that, in Holland at any rate, cervine fascioliasis is usually acute and with high mortality so that there is little opportunity for the parasites to reach maturity and disseminate eggs in any useful numbers. This genus can live as long as its host and hence even low grades of infection must be considered important in areas of communal grazing. It is fair to state, however, that one wild ruminant at any rate is considered to be at risk from domesticated stock. The wisent suffers severely from fascioliasis hepatica and has done since records began in 1873, many deaths in the old pre-1914 stocks of true wild wisent being attributable to the fluke. In Polish animals the overall infection rate at present is 53·8% and in the Bialowiecza reserve it is 100%. These animals are in an area where the endemicity in cattle is maximal and stringent precautions are taken to prevent the introduction of further contamination to the wisent (Drodz, 1967).

Paramphistomum cervi is ordinarily non-pathogenic and only in massive attack is there any risk of serious pathogenic effect. Heavy infections have, however, been found in elk and red deer in eastern Europe and in the Polish stocks of wisent although the disease in domesticated cattle in these regions is sporadic.

Adult cestodes are rarely of any pathogenic consequence although they are common in both wild and domesticated ruminants. In domesticated sheep both *Stilesia hepatica* and *Thysanosoma actinoides* may occur in the bile ducts, causing obstructive jaundice and condemnation of carcases, but it would be surprising if many of these infections were found to have originated from wild ruminants in view of the high endemicity in the sheep themselves. The only larval cestode which has been noted as pathogenic in wild cervids is *Cysticercus tenuicollis*. Reports of mature cysticerci as a cause of debilitation in cervids may be dismissed as presumptive but there are a number of documented cases of hepatitis cysticercosa resulting from mass intake of onchospheres.

ZOONOSES

Hydatid infection is one of the most important of the helminth zoonoses in northern regions where the wild ruminant is of significance as a source of *Echinococcus granulosus* infection for domesticated dogs and, hence, for hydatid in man himself. Pulmonary hydatidosis is a serious problem in the caribou-hunting Indians and Eskimos, the wolf-caribou sylvatic cycle being broken into when caribou offal—chiefly lungs—is fed raw to dogs. In northern Norway and in Finland there is a high incidence of human hydatidosis based on a dog-reindeer cycle. A limiting factor in the further intrusion of the infection into a pastoral cycle is that both of the true northern subspecies, *E. g. borealis* and *E. g. canadensis*, have low infectivity for sheep and cattle although they themselves can infect man. The available final hosts in these regions include the domestic dog, the coyote and the wolf (Smyth, 1964). The Arctic fox, although the major final host of *E. multilocularis* in the North, is refractory to infection with *E. granulosus*.

There is evidence that the coyote is not a very satisfactory final host and the prime wild final host appears to be the wolf. Thus, Sweatman (1952) found 42 out of 73 timber wolves infected in Ontario and Cowan (1948) noted that, although timber wolf in Jasper National Park harboured the parasite, it was absent from coyotes.

Similarly, although it is a much less important reservoir of infection for man than the caribou, the moose appears to be the prime intermediate host in northern North America. All the cervid species of the subcontinent are susceptible but the available comparative records show higher incidences in moose. Harper *et al.* (1955) found, in Saskatchewan, 2 of 20 white-tailed deer infected, 3 of 14 barren ground caribou but 29 of 96 moose. Sweatman (1952) in Ontario found one of 353 white-tailed deer but 17 of 29 moose infected. High incidences in moose have been recorded from other areas of Canada and Rausch (1952) considered that nearly every aged moose in southern Alaska was infected.

The higher incidence in moose is reflected in the incidence of adult parasites in wolves and Freeman *et al.* (1961) noted that *E. granulosus* was about twice as common in wolves in areas where moose were more numerous than white-tailed deer. Since older moose carry heavier infections as a result of their longer experience of encounter with onchospheres, the culling of these by wolves would present the greater likelihood of continuing and increasing the intensity of the sylvatic cycle of infection. Although Sweatman (1952) observed that, in Canada south of the St. Lawrence where there are no timber wolves the moose are not infected with hydatid, both Rausch (1952) and Ritcey and Edwards (1958) have found moose infected in areas with no wolves and conclude that there the cycle may be continued in dogs.

Hydatid infection is the only zoonosis of any importance in which man is at risk from wild ruminants. Fascioliasis in man almost invariably originates from contamination of cress beds by infected sheep and human dicrocoeliasis, although rare, usually may also be traced to sheep sources. (Many suspected human infections with this fluke, it should be remarked, are found to be pseudoparasitism, eggs appearing in the faeces after ingestion of infected liver.) Similarly, schistosomiasis japonicum in man has as its essential ruminant source domesticated species of cattle and buffalo.

CONCLUSIONS

It is worth repeating that, in any discussion of helminthiasis, distinction must be made between the presence of infection and the presence of overt disease.

Clearly, from the information presented here, there is not often new introduction of helminths from wild to domesticated ruminants for, in most cases, the majority of the wild ruminants' species are already there. The important question, in the general scheme, is whether the burdens in the wild animal are liable to augment substantially the existing populations in the domesticated.

Certainly in cervids the available data indicates that actual burdens of parasites, and in particular of trichostrongyloids, are usually low;

within the writer's experience they tend to be considerably lower than the numbers of equivalent parasites in domesticated stock. The contribution made by the cervid is therefore likely to be trivial. Although the reason for the low burdens in the wild may partly be due to dispersion on the available grazing, many localities exist where there is manifest overpopulation with, nevertheless, inconsiderable infections and there may be a case for an examination of the dynamics of the host-parasite relationship in these animals, in view of increasing interest in the wild ruminant as a food animal. The original concepts of game farming and of domestication of the eland, for example, have now been extended to include the musk ox, which is being experimentally herded for its wool and meat in America, and the European elk, which is being bred for meat, milk and as a beast of burden in Russia.

The host-parasite lists presented here are a useful indication that the majority of helminths of wild ruminants have domesticated species as their prime hosts and, on balance, although the point has not been made, the degree of exchangeability of populations is much greater from domesticated to wild ruminants than vice versa. Hence, the wild ruminant, in theory at least, is at greater risk than the domesticated in any encounter in communal grazing. The results of Longhurst and Douglas (1953) are a single example of the failure of the deer to withstand challenge by sheep parasites in spite of their apparently well-adjusted relationship to their own species. Although not germane to the present discussion it should be pointed out that tolerance of their parasites by wild cervids is an inherently unstable equilibrium and can be upset by privation or in the presence of disease or injury.

Again, in general terms, although there are exceptions, none of the wild ruminants is sufficiently numerous at present on the grazing territories of domesticated stock to represent a substantial risk. It is only when domesticated stock which have been on anthelmintic regimes are introduced to wild ruminant range that the reservoir of infection in these latter animals may be considered important as a nucleus for subsequent spread.

With certain individual parasites the risk is easily expressed. *Fascioloides magna* infection, originating from cervids and highly pathogenic for sheep, clearly presents a risk which may be calculated by graziers in endemic areas. With others the danger is almost impossible to assess, although it exists, as in the case of *Fasciola hepatica* infection. It may be suggested that, if this infection in domesticated stock is accepted by the stockowner without any attempt to apply prophylaxis, then the presence of similar levels of infection in the indigenous wild stock is largely cancelled out, as happens in areas of subsistence farming in many parts of the world. Where, however, the grazier is attempting control in his own stock, the reservoir of infection in the wild animals is a continuous economic threat. This is a field in which epidemiological research is required for there is no solution acceptable at present both to the stockowner and the conservationist.

If the truism may be forgiven, the greatest risk to a domesticated sheep or bovine is not a wild ruminant but another sheep or bovine. The most important reservoir of infection is in their own kind and, until prophylactic husbandry and therapy have been developed to a much greater extent, the rôle of the wild ruminant as a background of helminth infection is, in general, a minor one. What it is hoped has been indicated here is the almost insignificant amount of useful knowledge available when the ideal situation exists in domesticated stock and the problem of the wild animal and its infections must be approached.

REFERENCES

Alibasoglu, M., Kradel, D. C. and Dunne, H. W. (1961). Cerebral nematodiasis in Pennsylvania deer. (*Odocoileus virginianus*). *Cornell Vet.* **51**, 431–441.

Anderson, R. C. (1963a). Studies on *Pneumostrongylus tenuis* (Dougherty, 1945) of *Odocoileus virginianus*. *J. Parasit.* **49** (Supp.), 47.

Anderson, R. C. (1963b). The incidence, development and experimental transmission of *Pneumostrongylus tenuis* (Dougherty) (Metastrongyloidea: Protostrongylidae) of the white-tailed deer (*Odocoileus virginianus borealis*) in Ontario. *Can. J. Zool.* **41**, 775–792.

Anderson, R. C. (1964). Motor ataxia and paralysis in moose calves infected experimentally with *Pneumostrongylus tenuis* (Nematoda: Metastrongyloidea). *Northeastern Wildl. Conf.*, Hartford, Connecticut. January 1964.

Anderson, R. C. and Strelive, U. R. (1967). The penetration of *Pneumostrongylus tenuis* into the tissues of white-tailed deer. *Can. J. Zool.* **45**, 285–289.

Boev, S. N. (1957). The fauna of lung nematodes of ruminants of Kazakhstan and their adaptation to the hosts and environment. (Also Israel Programme for Scientific Translations. Jerusalem) *Zool. Zh.* **33**, 779–787.

Cameron, T. W. M. (1931). On two new species of nematodes from the Scottish red deer, *J. Helminth.* **9**, 213–216.

Campbell, W. C. and Todd, A. C. (1954). Natural infections of *Fascioloides magna* in Wisconsin sheep. *J. Parasit.* **40**, 100.

Cowan, I. M. (1948). The occurrance of the granular tapeworm *Echinococcus granulosus* in wild game in North America. *J. Wildl. Mgmt* **12**, 105–106.

Das, K. M. and Whitlock, J. H. (1960). Subspeciation in *Haemonchus contortus* (Rudolphi, 1803) Nematoda, Trichostrongyloidea. *Cornell Vet.* **50**, 182–197.

De Giusti, D. L. (1963). Incidence and distribution of *Elaphostrongylus odocoilei* in Michigan deer herd. *J. Parasit.* **49**, 47.

de Nahlik, A. J. (1959). Wild deer. Faber and Faber, London.

Drozdz, J. (1961). A study on helminths and helminthiases in bison, *Bison bonasus* (L.) in Poland. *Acta parasit. pol.* **9**, 55–95.

Drozdz, J. (1963). Helmintofauna zaaklimatyzowanego w Polsce Jelenia sika (*Cervus nippon L.*) *Wiad. parazyt.* **9**, 133–138.

Drozdz, J. (1967). The state of research on the helminthofauna of the European bison. *Acta theriol.* **12**, 377–384.

Dunn, A. M. (1965a). The gastro-intestinal helminths of wild ruminants in Britain. 1. Roe deer, *Capreolus capreolus capreolus. Parasitology* **55**, 739–745.

Dunn, A. M. (1965b). Parasites of red deer in Scotland. *Rep. Red Deer Commn* **1964**, 19–20.

Eveleth, D. F. and Bolin, F. M. (1955). Parasitic gastritis on elk. *J. Wildl. Mgmt* **19**, 152.

Freeman, R. S., Adorjan, A. and Pimlott, D. H. (1961). Cestodes of wolves, coyotes and coyote-dog hybrids in Ontario. *Can. J. Zool.* **39**, 527–532.

Grieder, H. (1934). Beobachtungen uber Rehkrankheiten in nordostschweizerischen Jagdrevieren. *Schweiz. Arch. Tierheilk.* **76**, 609–617.

Gushanskaja, L. Kh. (1934). [New nematode in sheep *Ostertagia spiculoptera*]: In 155–156. [*Trichostrongylidosis of ruminants*] by Skrjabin, K. I. and Orloff, I. V., Moscow.

Harper, T. A., Ruttan, R. A. and Benson, W. A. (1955). Hydatid disease (*Echinococcus granulosus*) in Saskatchewan big game. *Trans. N. Am. Wildl. Conf.* **20**, 198–208.

Herman, C. M. (1947). Further observations on deer foot worm infections. *Calif. Fish Game* **33**, 54.

Hübner, F. (1938). Die Magenwurmseuche des Rindes. Beobachtungen und Untersuchungen uber die Bekämpfungmöglichkeiten. *Dt. tierärztl. Wschr.* **46**, 143.

Jansen, J. Jr. (1958). Lebmaagtrichostrongyliden bij Nederlandse herten. Thesis. Univ. of Utrecht.

Jansen, J. Jr. (1960). Trichostrongylids in the fourth stomach of roe deer and red deer in the Netherlands. *Trans. Congr. Int. Union Game Biol.* **4**, 91–95.

Jansen, J. Jr. (1963). Some problems related to the parasite inter-relationships of deer and domestic animals. *Trans. Congr. Int. Union Game Biol.* **6**, 127–131.

Joyeux, Ch. and Gaud. I. (1946). Réchèrches helminthologiques marocains. *Archs. Inst. Pasteur Maroc.* **3**, 383–461.

Karabaev, D. K. (1953). [Changes in the helminth fauna of sheep imported into the Betpak-Dala desert plateau (Central Kazakhstan).] In *K. I. Skrjabin 75th Birthday Commemorative Volume*: 284–287. Moscow.

Karmanova, G. A. and Murtzaev, A. (1953). [The helminth fauna of *Gazella subgutturosa*.] In *K. I. Skrjabin 75th Birthday Commemorative Volume*: 288. Moscow.

Krems, J. (1939). Fallwilduntersuchungen 1935 mit 1938. *Tierarztl. Rdsch.* **45**, 762–766 and 773–776.

Leiper, J. W. G. and Clapham, P. A. (1938). Some nematode parasites found in Chinese water deer (*Hydropotes inermis*) with a description of *Trichostrongylus cervarius* n.sp. *J. Helminth.* **16**, 77–82.

Longhurst, W. M. and Douglas, J. R. (1953). Parasite interrelationships of domestic sheep and Columbian black-tailed deer. *Trans. N. Am. Wildl. Conf.* **18**, 168–188.

Longhurst, W. M., Leopold, A. S. and Dasman, R. F. (1952). A survey of California deer herds, their ranges and management problems. *Calif. Fish Game Biol.* No. 6.

Lucker, J. T. and Dikmans, G. (1945). The distribution of *Pseudostertagia bullosa* and some new records of nematodes from the pronghorn antelope (*Antilocapra americana*). *Proc. helm. Soc. Wash.* **12**, 2–4.

Mönnig, H. O. (1931). Wild antelopes as carriers of nematode parasites of domestic ruminants. *17th Rep. Div. Vet. Serv. S. Africa* 233–254.

Olsen, O. W. (1949). White-tailed deer as a reservoir host of the large American liver fluke. *Vet. Med.* **44**, 26–30.

Rausch, R. A. (1952). Hydatid disease in Boreal regions. *Arctic* **5**, 157–174.

Ritcey, R. W. and Edwards, R. Y. (1958). Diseases of the Wells Gray moose herd. *J. Mammal.* **39**, 139–145.

Roberts, F. H. S., Turner, H. N. and McKevett, M. (1954). On the specific distinctness of the ovine and bovine strains of *Haemonchus contortus* (Rudolphi) Cobb (Nematoda: Trichostrongylidae). *Aust. J. Zool.* **2**, 275–295.

Ronéus, O. and Nordkvist, J. (1962). Cerebrospinal and muscular nematodiasis (*Elaphostrongylus rangiferi*) in Swedish reindeer. *Acta vet. Scand.* **3**, 201–225.

Samson, K. W., Allen, R. W. and Schad, G. A. (1964). Comparative pathogenicity in *Ovis aries* of homologous and heterologous strains of *Haemonchus* (Nematoda: Trichostrongylidae) from domestic and wild sheep. *J. Parasit.* **50**, 421–426.

Skrjabin, K. I. (1931). [*The parasitic worms of the Arctic reindeer*] Moscow-Leningrad.

Smyth, J. D. (1964). The biology of the hydatid organisms. *Adv. Parasit.* **2**, 169–219.

Sweatman, G. K. (1952). Distribution and incidence of *Echinococcus granulosus* in man and other animals with special reference to Canada. *Can. J. Publ. Hlth* **43**, 480–486.

Whitlock, S. G. (1939). The prevalence of disease and parasites in white-tailed deer. *Trans. N. Am. Wildl. Conf.* **4**, 244–247.

Symp. zool. Soc. Lond. (1968) No. 24, 249–271.

MYCOTIC INFECTIONS

PETER K. C. AUSTWICK

Central Veterinary Laboratory, Weybridge, Surrey, England

SYNOPSIS

Mycotic infections in wild animals are perhaps the most neglected diseases of wildlife even though they were among the first microbial diseases to be recognized. Their importance in public and animal health and their relationship to the fluctuations in animal populations are now being investigated at an increasing rate. Two groups of mycoses can be recognized; the superficial which are almost always contagious and the systemic in which contagion plays only a minor rôle. Most of the fungi causing mycoses are now known to exist saprophytically in soil and in plant debris and infection may be derived partly or entirely from these sources.

Ringworm has been recorded in many wild animals, occasionally in the epidemic form, but most information on the occurrence of dermatophytes has come from surveys of trapped animals for unaffected "carriers" of species which infect man. Infection by the actinomycete *Dermatophilus congolensis* is slowly being recognized in wild animals and infections caused by *Candida* spp. are on record.

Of the non-contagious systemic mycoses, adiaspiromycosis is probably the commonest and most important especially in smaller burrowing rodents and insectivores. Aspergillosis causes a substantial proportion of diagnosed wild bird mortality, especially in freshwater and sea birds. Histoplasmosis and coccidioidomycosis in wild animals are of importance in human health by helping to demarcate the endemic areas of these diseases but they may also play a part in controlling populations. Cryptococcosis is worldwide but rarely diagnosed, despite the frequency of the causal fungus, *Cryptococcus neoformans*, in bird droppings. Sporotrichosis has been observed twice in rats and may be transmitted to man by bites. Penicillosis has been observed once in the bamboo rat. South American blastomycosis is at present exclusively a human disease but *Paracoccidioides brasiliensis* has now been recovered from bat intestines, implying a possible epidemiological relationship comparable to that already demonstrated between these animals and *Histoplasma capsulatum*.

The occurrence of mycotic infections in wildlife is now well established and the present need is for the assessment of their environmental rôle in animal populations. Their human and animal health implications have already led to much of our knowledge.

INTRODUCTION

Mycotic infections were among the first diseases of wildlife to be discovered and such was the interest shown that pathogenic fungi soon came to preoccupy those workers in human infectious disease in whom the idea of microbial origin was germinating. Aspergillosis was probably observed in a wild scaup duck (*Aythya marila*) as early as 1813 by Montague and in 1833, Richard Owen reported in the Proceedings of the Zoological Society of London, a case in a captive flamingo.

Rousseau and Serrurier saw what appears to have been aspergillotic infection in the lung of a deer (*Cervus axis*) in 1841 and in 1857 came the description of mouse favus by Gluge and d'Udekem in Belgium. Despite playing this vital early part in the formulation of the germ theory of disease, the subject was rapidly overtaken by the impetus given to bacteriology by the work of Pasteur and Koch and is only now emerging again as a recognized entity.

There are now some 25 mycoses of man and animals described and *c*. 200 fungi listed as their causal agents (see M.R.C. Memorandum No. 23, 1967). Many of these have been recorded in wild animals during the last 150 years but unfortunately few actual cases have been described. In fact the mycopathological part of wild animal disease remains in the case-collecting and descriptive phase, whilst the distribution of the mycoses still largely follows the distribution of interested workers. However, in recent years epidemiological studies have been undertaken in many countries and these may lead to a better understanding of incidence and ecological significance in these diseases. Already most of the fungi pathogenic to man and animals have been found to grow in the soil as saprophytes of specialized substrates, especially decaying keratinized animal tissue and dung, and the nutritional requirements which go with this specialization appear to enhance their pathogenic potential. The concept of the soil as the original source of the infectious propagules of pathogenic fungi is now widely accepted and a great many warm-blooded animals are now known to be susceptible to almost all known fungal infections. Even so, evolution has led to remarkable degrees of host specificity especially in the dermatophytes.

The most important distinction between the different mycoses is in their mode and ease of transmission. A few are highly contagious and are passed on from animal to animal without resort to a soil-inhabiting phase and in this category come ringworm and *Dermatophilus* infection. Most of them are essentially non-contagious and "soil-borne" in as much as the causal fungi grow in the soil or on plant debris and each animal acquires infection from these sources by its own activities and only very rarely becomes infected by fomites from other animals. In this account a number of the mycotic infections of free-living, wild mammals and birds are described and their epidemiology and effect on wildlife populations are discussed whenever sufficient information is available. A few of the examples quoted were specimens sent to Weybridge for examination but a number are drawn from the literature already well-documented by Halloran (1955), McDiarmid (1962), Vanbreuseghem (1963) and Smith and Austwick (1967).

CONTAGIOUS AND SUPERFICIAL MYCOSES

Ringworm

There are some 20 species of fungi capable of causing ringworm in man and animals by invading the keratinized layers of the skin, hair, feathers or horn. Their growth *in vivo* is characteristic, consisting of narrow, branching hyphae which eventually fragment to form arthrospores ranging from $1-10\,\mu$ in diameter and it is restricted to the keratinized tissues except where severe folliculitis causes a breakdown of the skin structure. In the soil and in culture, several other sporing stages are produced and it is now recognized that the dermatophytes are all Ascomycetes belonging to the family Gymnoascaceae. All these spore types from the micro- and macroconidia to the ascospores may be of significance in the transmission of the disease but possibly of greater importance in the carriage and distribution of these fungi among animals are the spiral hyphal appendages which are formed in nature at the periphery of both sexual and asexual reproductive structures. Although it is by no means certain that the function of these organs is concerned with distribution, the ease with which these spirals can become attached to the hairs of small rodents and insectivores could be an important epidemiological factor.

Although there are now numerous records of the isolation of ringworm fungi from free-living wild animals, very few are from actual lesions and in many of the surveys so far conducted, microscopical examination of skin scrapings and hairs has not always been carried out. In Table I some of the available information on the incidence of infection and of recoveries in culture is summarized, showing the wide range of these percentages. It seems that only intensive surveys in small areas, such as that by English and Southern (1967), can give figures which reflect true incidences and provide a basis for the comparison of the infection rates in different species of animal. The larger scale surveys certainly demonstrate that small, burrowing species of mammals may play an important ecological rôle in the distribution of ringworm fungi.

Mammals

The classical wild animal mycosis is mouse favus caused by *Trichophyton mentagrophytes* var. *quinckeanum*. As a disease it is spectacular, usually fatal and is transmissible to other animals and to man. It occurs sporadically in almost all temperate regions but occasionally reaches epidemic proportions especially when the mouse population

TABLE I

Percentage of ringworm infected and "carrier" mammals

Host species	Dermatophyte	Size of sample	% with lesions	% micro-positive	% culture positive	Location	Reference
21 species (averages only)	M. cookei, M. gypseum, T. mentagrophytes	1142	0	0	7·7	S. W. Georgia, U.S.A.	Menges et al. (1957)
12 species (averages only)	M. gypseum, T. terrestre	605	0	—	5·6 / 11·4	S. E. Moravia	Očenašek and Dvořák (1962)
11 species (averages only)	T. mentagrophytes, T. terrestre	218	0	—	4·1	Roumania	Alteras et al. (1966)
12 species (averages only)	T. mentagrophytes, M. gypseum, M. cookei	904	0	0	2·5 / 9·0	Slovakia	Chmel and Buchvald (1967)
Ondatra zibethicus (muskrat)	T. mentagrophytes	364	9·6[1]	—	—	Iowa, U.S.A.	Errington (1942)
Lepus americana (snow shoe hare)	T. mentagrophytes	—	35·0	—	—	Montana, U.S.A.	Adams et al. (1956)
Mus musculus (house mouse)	T. mentagrophytes	308	—	—	3·6	Egypt	Taylor et al. (1964)
Apodemus sylvaticus (wood mouse)	T. mentagrophytes	48	0	—	10·4	England	English (1967a)[2]
Erinaceus europaeus (hedgehog)	T. mentagrophytes var. erinacei	114	0	—[3]	44·7	South Island New Zealand	Smith and Marples (1963)
		117	0	—	47·9	North Island New Zealand	English et al. (1964)
		50	0	28·0	—	New Zealand	
Clethrionomys glareolus (bank vole)	T. persicolor	127	0	—	53·0	England	English and Southern (1967)
Microtus agrestis (short-tailed vole)		65	1·5	—	40·0		English (1967b)
		113	0	—	25·0		English and Southern (1967)
Apodemus sylvaticus		26	0	—	19·0		English and Southern (1967)

Host species	Fungus	No. examined			%	Location	Reference
Tatera indica (Indian gerbil)	*Arthroderma simii* (*Trichophyton simii*)	170	—	—	11·8	India	Gugnani *et al.* (1967)
Suncus murinus (shrew)		62	—	—	12·9		
Mus musculus	*T. mentagrophytes*	243	0	0	1·2	S.W. Georgia, U.S.A.	McKeever *et al.* (1958b)
Peromyscus gossypinus (cotton mouse)	*M. gypseum*	58	0	0	34·5		
Sigmodon hispidus (cotton rat)	*T. mentagrophytes*	351	0	0	16·2		
Peromyscus polionotus (old-field mouse)	*M. gypseum* / *M. cookei*	311	0	0	9·3		
Rattus norvegicus (common rat)	*T. mentagrophytes*	350	0	0	4·9	S.W. Georgia, U.S.A.	Smith *et al.* (1957)
Rattus rattus (black rat)	*M. gypseum*	84	0	0	7·1		
Didelphis marsupialis (opossum)	*T. mentagrophytes* / *M. gypseum*	436	0	0	3·9	S.W. Georgia, U.S.A.	McKeever *et al.* (1958a)
Sylvilagus floridanus (cottontail rabbit)		207	0	0	0·5		
Procyon lotor (raccoon)	*M. cookei*	583	0	0	0·1		
Mephitis mephitis (striped skunk)		239	0	0	0·2		

[1] Of litters.
[2] All 5 positive mice from one indoor habitat.
[3] "Limited number examined and found microscopically positive."
M. = *Microsporum*.
T. = *Trichophyton*.

rises, e.g. during the "mouse plague years" of 1917 and 1952 in Australia (Elton, 1931; Donald and Brown, 1964). The term "favus" refers to the clinical appearance of the lesions as a honey-combed area of skin which exudes serum drops and generally a large area of the anterior body surface is affected. The lesion is basically a folliculitis in which vast numbers of arthrospores are formed in the cup-shaped scabs. Transmission occurs to predatory cats and dogs and to human beings who handle contaminated straw, etc.

Whereas *T. mentagrophytes* var. *quinckeanum* is rarely reported, the type variety *T. m.* var. *mentagrophytes* is common everywhere and appears in most of the surveys listed in Table I. Its association with mice caught in human habitations represents an important relationship between wildlife and public health. Several other species of wild animal have been found to be infected or to carry this species and during a coypu (*Myocaster coypus*) eradication campaign in East Anglia, we recovered this species from muzzle lesions on one animal and also isolated it from three others, out of a total of 19, which did not show lesions. The material was kindly sent to us from Norwich by Dr. R. M. Newson and Mr. R. G. Holmes.

T. mentagrophytes var. *erinacei* occupies a unique position in its specificity to the hedgehog (*Erinaceus europaeus*) and has been studied in New Zealand by Smith and Marples (1963) and by English *et al.* (1964). Lesions, when seen, mostly occur in the nose and ears as crusty scaling scabs and the fungus is distinguished in culture by the production of a yellow pigment, as distinct from the common reddish one of most other varieties. A high percentage of actual infection is a characteristic of the hedgehog, already noted for its arthropod parasite load, and the characters of the fungus enable the investigation of epidemiological links of similar isolates obtained from man and domesticated animals (English *et al.*, 1962).

T. persicolor was once regarded as a variety of *T. mentagrophytes* but has recently been shown to have a distinct perfect state, *Nannizzia persicolor* Stockdale. It had never been found infecting wild animals until English (1966) isolated it from a lesion on the tail of a vole (*Microtus agrestis*). During further studies of these small rodents (English and Southern, 1967) the fungus was recovered from a relatively high proportion of this species and of the bank vole (*Clethrionomys glareolus*) although no lesions were apparent on the skin. These high percentages indicate that almost all individuals of these species must carry the fungus at some time during their life.

T. verrucosum, the cause of cattle ringworm, has always failed to reveal infections in wild animals which might serve to spread the disease from farm to farm but recently Koch (1965) has isolated this species from skin lesions on a mole (*Talpa europaea*).

Birds

Ringworm infection in wild birds presents an unsolved problem both in its occurrence and its causal agents. The only authentic record is one of *Trichophyton gallinae* infection in a black grouse (*Lyrurus tetrix*) by Pätiälä (1951) in Finland but reports of "bald-headed" birds are becoming more frequent. Since 1954 material from 10 affected wild birds has been sent to Weybridge by Dr. D. K. Blackmore and Dr. J. W. Macdonald and on 8 occasions hyphae were found in the skin lesions. These were in 4 robins (*Erithacus rubecula*), a blue-tit (*Parus caeruleus*), a bullfinch (*Pyrrhula pyrrhula*), a house sparrow (*Passer domesticus*) and a hedge sparrow (*Prunella modularis*). The fungal isolates obtained in 6 cases closely resembled dermatophytes in their gross appearance in culture but failed to produce the characteristic spores and in two isolates inoculated on to chicken, failed to induce lesions.

Epidemiology

By parallel human and animal studies, Chmel (1967) and his colleagues in Slovakia have shown how the percentage of human ringworm infection due to *Trichophyton verrucosum* and to *T. mentagrophytes* differ widely in their geographical distribution. In Southern and lowland Slovakia the farms are chiefly large cooperatives, each with many head of cattle, and the environment is therefore very favourable for the rapid spread of *T. verrucosum* infection. This species then accounts for most of the human ringworm in the area. In the mountainous areas in Northern Slovakia, where cattle are housed in small groups, the predominant human infection is caused by *T. mentagrophytes*, which is thought to be derived from rodents living in the hay and straw of the barns. This view is apparently confirmed by the seasonal distribution of cases with the peak of *T. verrucosum* infection occurring in January but that of *T. mentagrophytes* in the spring, with the clearing of the lowest layers of the stored food and straw, i.e. those most likely to be harbouring the rodent population.

With the emphasis on the public health aspects of the wild animal "reservoir" of ringworm in recent years, the effect of infection on the animals themselves has been largely overlooked. From the literature this would seem negligible but there is some evidence that at least epidemic ringworm can be associated with extensive mortality in rodents, e.g. in the mouse favus in Australia (*l.c.*) and in the muskrat

(*Ondatra zibethicus*) outbreaks reported in Iowa by Errington (1942). The latter is a detailed population study in which 9·6% of 364 litters were found infected with *T. mentagrophytes* which was thought to be responsible for a mortality rising each season from 2·8% in litters born in April–May to 26·7% in those born in August. This is probably the first evidence of population control by a fungal disease, even if only effective in the later litters, and it seems likely that other examples will be discovered in the future.

Dermatophilus *infection*

The actinomycete *Dermatophilus congolensis* causes one of the commonest skin diseases of domesticated animals and yet few cases have ever been reported in wild animals. The distribution of the infection among cattle, sheep and horses is world-wide. Reports from tropical Africa indicate that the giraffe (*Giraffa cameleopardalis*), Thomson's gazelle (*Gazella thomsonii*) and a zebra (*Equus* sp.) have had lesions observed on them in the wild (Mackenzie and Simpson, 1956; Green, 1960) but there is no evidence that it causes the severe and fatal disease (streptotrichosis) that is frequently seen in domesticated stock. In temperate regions infection of the legs has been reported in a white-tailed deer (*Odocoileus virginianus*) in New York State by Dean *et al.* (1961) and on the nose in two young chamois (*Rupicapra rupicapra*) in Switzerland by Nicolet *et al.* (1967). Recently a fox (*Vulpes vulpes*) was brought to the Central Veterinary Laboratory at Weybridge and my colleague Mr. G. A. Pepin found typical filaments of *Dermatophilus* in crust-like lesions on the ears. This seems to be the first record of the disease in a carnivore. It can hardly be deduced from these few records whether there is transmission of infection between wild and domesticated animals but direct or indirect contact during wet weather would probably be the only necessary factor for the transfer of the motile zoospores.

Candida *infection*

Yeasts and bacteria may be equally ubiquitous as commensals of the skin and alimentary tracts of warm-blooded animals and many yeast species have been isolated from free-living wild birds and mammals (Van Uden, 1963). These include *Candida albicans* the cause of candidiasis and it is remarkable that this infection has been recorded so few times in the wild (McDiarmid, 1962). This fact led Vanbreuseghem (1963) to say that it is a "disease of the civilized animal". Inadequate levels of diet are often associated with outbreaks of candidiasis in domesticated stock and also in artificially reared game birds (Keymer and Austwick, 1961) but it seems likely that the under-nourished wild animal falls victim to predator or carrion-eater long before it reaches the post-mortem table.

In 1966 a scaup duck (*Aythya marila*) was found in Lincolnshire and from lesions on the head Mr. M. G. Jones isolated *Candida albicans*. Through the kindness of Dr. A. Apinis, I examined the preserved head and found heavily encrusted lesions all around the base of the bill and over the top of the head. Within the upper mandible were a number of 3 mm diam. plaques of desquamated epidermal cells and in all the affected tissue the pseudohyphae and cells of the yeast were detectable. Sections of the head lesion showed a massive proliferative dermatitis with the invasion of the dermis, stratum corneum and bases of the feathers. It seems likely that the lesions were of long-standing and that the heavy crusting over the head, possibly causing blindness, had contributed to the death of the bird.

Evidence for the contagious nature of *Candida albicans* infection is hard to find but the epidemiology of the outbreaks of the disease in poultry has indicated rapid spread. One important and as yet un-assessed factor may be the presence of *C. albicans* on herbage in moist summer weather, as reported by Keymer and Austwick (1961).

NON-CONTAGIOUS AND SYSTEMIC MYCOSES

Adiaspiromycosis (haplomycosis)

By far the best known and probably most widespread systemic mycosis of wild animals is adiaspiromycosis caused by the two *Emmonsia* spp., *E. parva* and *E. crescens*. It may play a significant rôle in the control of populations in certain seasons. Presumably infection occurs by the inhalation of the small, 2–4 μ diam., conidia produced by the mycelium growing in the soil but only one instance of soil isolation can be cited (Ciferri and Montemartini, 1959). The number of conidia inhaled appear to be directly related to the number of lesions observed per lung, for each spore simply enlarges and remains *in situ*. These spherules reach a maximum of 14 μ in *E. parva* but are generally larger in *E. crescens* reaching to 500 μ with a wall of 50 μ thickness. Reaction of the lung tissue is often minimal but with such a large expanding body, collapse of the adjacent alveoli occurs with infiltration and epithelialization of the surrounding tissues. Recent reviews list up to 50 species of host mammal, most of which live in burrows (Dvořák et al., 1966; Emmons and Jellison, 1960).

TABLE II

Percentages of animals infected by Emmonsia crescens (*as well as* E. parva, *marked**)

Host species	Country	Month captured	Percent- age infected	Reference
Apodemus flavicollis	Czecho- slovakia	Jan.–Apr.	5·4	Prokopič et al., 1965
Apodemus flavicollis	Czecho- slovakia	May	7·5	Dvořák et al., 1967
Apodemus flavicollis	Sweden	winter	3·2	Jellison, 1956
Clethrionomys glareolus	Czecho- slovakia	Jan.–Apr.	4·6	Prokopič, et al., 1965
Clethrionomys glareolus	Czecho- slovakia	May	12·5	Dvořák et al., 1967
Microtus arvalis	Czecho- slovakia	Jan.–Apr.	1·4	Prokopič et al., 1965
Microtus arvalis	Czecho- slovakia	April	10·0	Dvořák et al., 1967
Myocastor coypus	England	mostly winter	12·8	
Talpa europaea	England	May	80·6	McDiarmid and Austwick, 1954
Rodentia and Insectivora	Arizona*	—	33·3	Ashburn and Emmons, 1942
Rodentia and Insectivora	Alberta	—	5·1	Dowding, 1947
Rodentia and Insectivora	Korea	autumn	2·5	Jellison, 1954
Rodentia and Insectivora	Norway	—	1·0	Jellison et al., 1960
Rodentia and Insectivora	Sweden	summer	3·3	Paldrok and Zetterberg, 1962

Recent records from the United Kingdom have added three more host species to the list—hedgehog (*Erinaceus europaeus*) in which three out of nine examined by Dr. A. McDiarmid showed spherules in their lungs, two of them with such large numbers that the lung function must have been severely impaired; coypu (*Myocastor coypus*) in which a survey of 70 lungs carried out in East Anglia in conjunction with Dr. E. A. Gibson revealed 9 infected animals (12·6%) and rabbit (*Oryctolagus cuniculus*) in which Dr. A. McDiarmid detected three cases. One was of particular interest because as well as lesions containing single spherules there were many others containing up to 8 (seen in cross section). Many of these spherules showed signs of disintegration, firstly, by the rupture and collapse of the thick wall often with penetration of neutrophils into the interior and, secondly, by the separation of the wall into several lamellae.

Information on the incidence of *Emmonsia* infection is sparse but has been summarized in Table II. This shows the wide range of percentages of animals found infected during surveys in different parts of the world and emphasizes that the Spring peak noted by Dvořák *et al.* (1967) in Czechoslovakia probably also occurs in other countries. Whether the present data on adiaspiromycosis allows the assumption that it is responsible for controlling wild animal populations in any way is hard to say. It seems likely that a combination of factors including high population density, other types of disease and reduced food supply contributes to the widespread springtime mortality which occurs in certain years. In particular there is need for a closer examination of the factors favouring the rapid growth of *Emmonsia* spp. in soil.

Aspergillosis

The early recognition of aspergillosis in wild birds has been mentioned in the introduction and it is now widely diagnosed, judging from reports on wildlife disease throughout the world. The chief cause is *Aspergillus fumigatus* but *A. flavus* and *A. nidulans* have both been isolated from wild birds. Aspergillosis has been reported in many species of birds from very different habitats, chiefly in the temperate zones, but as yet no indigenous case in a penguin has been found despite the great susceptibility of these birds in captivity. As in domesticated poultry, the disease occurs in both sporadic and epidemic forms, often as chronic air-sac infection. Sea and lakeside strand debris in which *A. fumigatus* has grown has been considered responsible for several outbreaks notably the one among herring gulls (*Larus argentatus*) in Boston Harbour described by Davis and McClung (1940). Few data are available on the incidence of aspergillosis in wild bird populations. Figures from general surveys of bird mortality are generally very low, e.g. 7 cases out of 460 birds examined by Keymer (1958) and 18 in a 10-year period by McDiarmid (1955), but Poulding (1957) found the disease in 14 out of 38 herring gulls and considered it was one of the major causes of death in these birds. Heavy losses in mallards (*Anas platyrhynchus*) were reported by Neff (1955) when the birds had fed on mouldy silage during a period of very cold weather.

The theory of the derivation of infection from concentrated point sources of spores is to some extent confirmed when the numbers of spores required to set up infection is considered. Herman and Sladen (1958) found that 10–50 million were necessary by the intra-thoracic route in young chicken, whilst Dr. E. C. Appleby and I have been able to estimate that at least 5 million inhaled spores were needed to produce fatal infection in day-old chicks and at least 17 million in adult birds.

Pulmonary aspergillosis in wild mammals has rarely been reported. Hülphers and Lilleengen (1947) saw the infection in hares (*Lepus europaeus*) in Norway and a characteristic form has been observed by Burgisser (1955) in roe-deer (*Capreolus capreolus*) in Switzerland. This consisted of a tumour-like lesion in the conch and ethmoid regions in five animals, with typical miliary lung lesions in three of them. *A. fumigatus* was isolated in one case. Early in 1966 Dr. A. McDiarmid found a mole (*Talpa europaea*) with a large nodular lesion in the lungs and isolated *A. fumigatus* from it. Histopathological examination showed extensive pulmonary congestion with copious bronchiolar exudate containing individuals and eggs of a lungworm. The main lesion consisted of numerous foci of infection within alveoli, in which swollen-celled hypae were surrounded by an intensive neutrophilic infiltration. This microscopical appearance is similar to that seen in acute aspergillosis in lambs and it is thought that this mole must have inhaled a large number of *A. fumigatus* spores a short time before its death.

Histoplasmosis

Histoplasmosis is a widespread infection of man and animals in the warmer parts of the world and the causal organism *Histoplasma capsulatum* has become recognized in recent years as one of the most important pulmonary pathogens of man in the U.S.A. Emmons and Ashburn (1948) were first to demonstrate infection in wild *Rattus norvegicus* and at the time this was thought to be of direct epidemiological significance in maintaining a reservior of infection for man. Wild animal infection is now, however, regarded as a useful indicator of the presence of *H. capsulatum* in an area in which all animals are exposed to infection from the common source in the soil.

Most records of this fungus in wild animals have come from studies carried out by the inoculation of the triturated organs of trapped animals into laboratory mice but actual infections have been confirmed in several species of rodent in the U.S.A. by Emmons *et al.* (1955). Subsequent reports of the association of histoplasmosis in man with caving excursions led to reports of the isolation of *H. capsulatum* from various organs of bats present in caves but direct infection has only rarely been demonstrated in these mammals (Klite and Diercks, 1965; Emmons *et al.*, 1966). The occurrence of a fungus strongly resembling *H. capsulatum* in a sub-mandibular lymph node of a badger (*Meles meles*) in Switzerland (Burgisser *et al.*, 1961) poses an epidemiological problem in an apparently non-endemic area for the fungus, while the appearance of infection by the other species, *H. duboisii*, in captive monkeys originating from Central Africa, may itself show that infection of the wild animal may also be occurring with this species (Mariat and Segretain, 1956; Walker and Spooner, 1960). Vanbreuseghem (1963) has called attention to what is perhaps the earliest recognition of *H. duboisii* in animals by Hamerton (1937) who found "Blastomycoides immitis" in a skin lesion of a guinea baboon (*Papio papio*) at the London Zoo. A glimpse at one of the original sections of this case shows the typical large intracellular yeast-phase cells of this fungus.

Other systemic mycoses

Coccidioidomycosis

Ashburn and Emmons (1942) first found *Coccidioides immitis* in the lungs of wild rodent when examining animals trapped in Arizona. The affected animals had typical endosporulating spherules 5–30 μ diam. enclosed in granulomatous lesions which were not considered to be of the progressive type. Subsequently the epidemiology has been thoroughly investigated (Maddy, 1960) and in trapping surveys in the endemic regions of the U.S.A., pocket mice (*Perognathus* spp.) and the kangaroo rat (*Dipodomys merriami*) have been consistently found to be infected (Swatek and Plunkett, 1957). The only record of animal infection outside the Americas is that of Kemenes (1954) in a wild rabbit in Hungary. The identity of the isolate was confirmed by Dr. G. A. deVries.

Cryptococcosis

Cryptococcosis is a worldwide disease of man and animals which remains relatively rare in spite of an apparently generalized exposure to infection in many environments. The main epidemiological advance in this disease was made when Emmons (1951, 1955) showed the prevalence of the causal organism, *Cryptococcus neoformans*, in bird droppings, especially those of pigeons. Staib (1962) has shown that this yeast possesses an advantage over its microbial competitors in this habitat by being able to utilize uric acid and creatinine for growth.

There is one wild animal in which cryptococcosis could be of great importance as a cause of death and this is the koala (*Phascolarctos cinereus*) for Backhouse and Bolliger (1961) have described one case in an animal found dying in the bush. Because some 5 out of a total of 23 examined (mostly from the Taronga Zoological Park in Sydney) showed this infection, these authors have suggested that cryptococcosis may play a significant rôle in the recent decline of the koala population.

Penicillosis

A curious disease of the bamboo rat (*Rhizomys sinensis*) discovered by Capponi *et al.* (1956) deserves mention although it was only observed in newly captured specimens. The causal *Penicillium*, *P. marneffei*, differs from other species of this genus in its pathogenicity and by growing intracellularly in macrophages as short mycelial branches which fragment to form arthrospores. These spores closely resemble the yeast phase of *Histoplasma capsulatum*. The three animals affected had extensive visceral lesions from which the fungus was readily isolated and used to produce similar experimental infections. No further cases have been reported.

South American blastomycosis

Paracoccidioides brasiliensis has remained exclusively a human pathogen but the recent recovery from the intestines of three bats (*Artibeus lituratus*) in Columbia by Grose and Tamsitt (1965) introduces a possible epidemiological relationship of the type already observed between bats and *Histoplasma capsulatum*.

Sporotrichosis

From first being described in wild rats (*Rattus norvegicus*) during a survey for plague infection by Lutz and Splendore (1907) in Brazil, sporotrichosis has been recognized as a human disease all over the world. Although usually rare, it is sometimes epidemic. The causal fungus, *Sporothrix* (*Sporotrichum*) *schenckii*, characteristically produces ulcerative lesions in the subcutaneous lymph nodes of the limbs and tail in which the cigar-shaped yeast phase cells 3–5 × 2–4 μ may be found with difficulty. Generalized infection was reported in a rat by Pringault and Vigne (1921). Transmission to man by rat and mouse bites has occurred but the mechanical introduction of the fungus into the wound from soil cannot be ruled out.

CONCLUSION

The study of the diseases among wild animals is beset from its outset by inherent difficulties in obtaining suitable material for examination. Our knowledge has come by two routes: by surveys of trapped or shot animals which provide local data and by the chance collection of diseased animals which gives some idea of the geographical distribution of the diseases. In the last ten years information on mycotic infections from these two sources has at least given sufficient records to show that wild animals are prone to the same mycoses that affect man and his domesticated stock and that they may play an important part in epidemiological pattern. It also enables the distinction to be made between the universal occurrence of many of the pathogenic fungi and the existence of specialized ecological nitches from which some species rarely emerge, e.g. *T. persicolor* on the bank and short-tailed vole and *T. mentagrophytes* v. *erinacei* on the hedgehog.

The present situation highlights the need for studies on the relationship between the fluctuation in animal numbers and the changes in the infection and "carrier" rates. With adiaspiromycosis, for instance, a causal relationship may operate in those animals with heavy and apparently fatal lung infections. Investigation of this would require quantitative recovery of the *Emmonsia* sp. from soil over several seasons and the correlation of this with the degree of infection observed. Similarly, *Aspergillus fumigatus* occasionally causes epidemics of aspergillosis in free-living birds which may again be related to the occurrence of conditions favourable for fungal growth in the offending substrate.

Not the least of the problems concerning wildlife mycoses is the danger of importing non-indigenous pathogens in captured animals intended for zoological collections. Such infections may not show clinically for some time after the animal has settled in captivity, such as has happened with *Histoplasma duboisii* in baboons on the three occasions mentioned. The danger does not only remain here for there is a very real hazard to laboratory workers who may isolate fungi, e.g. *Histoplasma capsulatum* and *Coccidioides immitis* from animal organs, and who may not be familiar with the extreme ease of acquiring infection from the airborne spores from cultures.

Besides infection there are two other types of fungal disease, namely allergy and toxicosis. The former is only really defined in man and it seems doubtful if it could play a detectable rôle in wildlife disease. The mycotoxicological aspect, on the other hand, could perhaps provide a long awaited clue to the epidemiology of human visceral cancer in those tropical countries where stored grain is regularly exposed to spoilage by moulds. Several metabolic products of these fungi, including aflatoxins from *Aspergillus flavus*, are already known to be carcinogenic to laboratory animals and a study of the liver lesions in rodents caught in and around human habitation in these areas might be rewarding.

It is clear that the mycoses discussed in this account are closely tied with all other aspects of wildlife biology and must be considered as an essential part of the environmental pressures outlined in the other contributions to this Symposium.

ACKNOWLEDGEMENTS

I would like to acknowledge the kind help of Miss M. P. English and Dr. A. McDiarmid in the preparation of this paper. Dr. I. F. Keymer kindly allowed me to examine Hamerton's slides of the skin of *Papio papio*.

REFERENCES

Adams, L., Salvin, S. B., and Hadlow, W. J. (1956). Ringworm in a population of snowshoe hares. *J. Mammal.* **37**, 94–99.

Alteras, I., Nesterov, V., and Ciolofan, I. (1966). The occurrence of dermatophytes in wild animals from Roumania. *Sabouraudia* **4**, 215–218.

Ashburn, L. L. and Emmons, C. W. (1942). Spontaneous coccidioidal granuloma in the lungs of wild rodents. *Archs Path.* **34**, 791–800.

Backhouse, T. C. and Bolliger, A. (1961). Morbidity and mortality in the Koala (*Phascolarctos cinereus*). *Aust. J. Zool.* **9**, 24–37.

Burgisser, H. (1955). Mycoses nasales chez le Chevreuil. *Schweizer. Arch. Tierheilk.* **97**, 434–438.

Burgisser, H., Fankhauser, R., Kaplan, W., Klinger, K. and Scholer, H. J. (1961). Mykose bei einem Dachs in der Schweiz. Histologisch Histoplasmose. *Pathologia Microbiol.* **24**, 794–802.

Capponi, M., Sureau, P. and Segretain, G. (1956). Penicillose de *Rhizomys sinensis*. *Bull. Soc. Path. exot.* **49**, 418–421.

Chmel, L. (1967). Epidemiological aspects of zoophilic dermatomycoses. *In* "Recent Advances in Human and Animal Mycology" pp. 21–36, L. Chmel, ed., Slovak Academy of Sciences, Bratislava.

Chmel, L. and Buchvald, J. (1967). Small rodents as reservoirs of trichophytoses. *In* "Recent Advances in Human and Animal Mycology". pp. 121–126. L. Chmel, ed., Slovak Academy of Sciences, Bratislava.

Ciferri, R. and Montemartini, A. (1959). Taxonomy of *Haplosporangium parvum*. *Mycopath. Mycol. appl.* **10**, 303–316.

Davis, W. A. and McClung, L. S. (1940). Aspergillosis in wild herring gulls. *J. Bact.* **40**, 321–323.

Dean, D. J., Gordon, M. A., Severinghaus, C. W., Kroll, E. T. and Reilly, J. R. (1961). Streptothricosis: a new zoonotic disease. *N.Y. St. J. Med.* **61**, 1283–1287.

Donald, G. F. and Brown, G. (1964). *T. mentagrophytes* and *T. mentagrophytes* var. *quinckeanum* infections in South Australian mice. *Aust. J. Derm.* **7**, 133–140.

Dowding, E. S. (1947). *Haplosporangium* in Canadian rodents. *Mycologia* **39**, 372–373.

Dvořák, J., Otčenášek, M. and Rosický, B. (1966). Conception on the circulation of *Emmonsia crescens* Emmons and Jellison, 1960 in Nature. *Folia parasit.* **13**, 150–157.

Dvořák, J., Otčenášek, M. and Prokopič, J. (1967). Seasonal incidence of adiaspores of *Emmonsia crescens* Emmons and Jellison 1960 in wildly living animals. *Mycopath. Mycol. appl.* **31**, 71–73.

Elton, C. (1931). The study of epidemic diseases among wild animals. *J. Hyg., Camb.* **31**, 435–456.

Emmons, C. W. (1951). Isolation of *Cryptococcus neoformans* from soil. *J. Bact.* **62**, 685–690.

Emmons, C. W. (1955). Saprophytic sources of *Cryptococcus neoformans* associated with the pigeon (*Columba livia*). *Am. J. Hyg.* **62**, 227–232.

Emmons, C. W. and Ashburn, L. L. (1948). Histoplasmosis in wild rats. Occurrence and histopathology. *Publ. Hlth Rep., Wash.* **63**, 1416–1422.

Emmons, C. W. and Jellison, W. L. (1960). *Emmonsia crescens* sp. n. and adiaspiromycosis (haplomycosis) in mammals. *Ann. N.Y. Acad. Sci.* **89**, 91–101.

Emmons, C. W., Klite, P. B., Baer, G. M. and Hill, W. B. (1966). Isolation of *Histoplasma capsulatum* from bats in the United States. *Am. J. Epid.* **84**, 103–109.

Emmons, C. W., Rowley, D. A., Olson, B. J., Mattern, C. F. T., Bell, J. A., Powell, E., and Marcey, E. A. (1955). Histoplasmosis. Proved occurrence of inapparent infection in dogs, cats, and other animals. *Am. J. Hyg.* **61**, 40–44.

English, M. P. (1966). *Trichophyton persicolor* infection in the field vole and pipistrelle bat. *Sabouraudia* **4**, 219–222.

English, M. P. (1967a). Notes on British mammals. No. 15. Ringworm in wild mammals. *J. Zool., Lond.* **153**, 556–561.

English, M. P. (1967b). The nature of *Trichophyton persicolor* infection in the bank vole and the interpretation of the results of sampling techniques. *Sabouraudia* **5**, 295–301.

English, M. P., Evans, C. D., Hewitt, M. and Warin, R. P. (1962). "Hedgehog ringworm". *Br. med. J.* **1962i**, 149–151.

English, M. P., Smith, J. M. B. and Rush-Munro, F. M. (1964). Hedgehog ringworm in the North Island of New Zealand. *N.Z. med. J.* **63**, 40–42.

English, M. P., and Southern, H. N. (1967). *Trichophyton persicolor* infection in a population of small wild mammals. *Sabouraudia* **5**, 302–309.

Errington, P. L. (1942). Observations on a fungus skin disease of Iowa muskrats. *Am. J. vet. Res.* **3**, 195–201.

Gluge, G. and d'Udekem, J. (1857). De quelques parasites végétaux développées sur des animaux vivants. *Bull. Acad. r. Belg. Cl. Sci.* **26**, 338–352.

Green, H. F. (1960). Streptothricosis in zebra and donkeys and demodectic mange in eland in Kenya. (letter). *Vet. Rec.* **72**, 1098.

Grose, E. and Tamsitt, J. R. (1965). *Paracoccidioides brasiliensis* recovered from the intestinal tract of three bats (*Artibeus lituratus*) in Columbia, S.A. *Sabouraudia* **4**, 124–125.

Gugnani, H. C., Shrivastav, J. B. and Gupta, N. P. (1967). Occurrence of *Arthroderma simii* in soil and on hair of small mammals. *Sabouraudia* **6**, 77–80.

Halloran, P. O'C. (1955). A bibliography of references to diseases in wild mammals and birds. *Am. J. Vet. Res.* **16**, 1–465.

Hamerton, A. E. (1937). Report on the deaths occurring in the society's gardens during the year 1936. *Proc. zool. Soc. Lond.* **107**(B), 443–474.

Herman, C. M. and Sladen, W. J. L. (1958). Aspergillosis in waterfowl. *Trans. N. Am. Wildl. Conf.* **23**, 187–191.

Hülphers, G. and Lilleengen, K. (1947). Mögelsvampsinfektion, aspergillos, hos hare och vilt levande Fågel. *Svensk Vet. Tidskr.* **52**, 235–239, 243–262.

Jellison, W. L. (1954). The presence of a pulmonary fungus in Korean rodents. *Publ. Hlth Rep., Wash.* **65**, 1057–1063.

Jellison, W. L. (1956). Haplomycosis in Sweden. *Nord. vet. Med.* **8**, 504–506.

Jellison, W. L., Vinson, J. W. and Holager, E. (1960). Haplomycosis in Norway. *Acta path. microbiol. scand.* **49**, 480–484.

Kemenes, F. (1954). Über einen falle von Coccidioidomykose bei einem Kaninchen in Ungarn. *Acta microbiol., hung.* **2**, 191–194.

Keymer, I. F. (1958). A survey and review of the causes of mortality in British birds and the significance of wild birds as disseminators of disease. Pt. I. *Vet. Rec.* **70**, 713–720.

Keymer, I. F. and Austwick, P. K. C. (1961). Moniliasis in Partridges (*Perdix perdix*). *Sabouraudia* **1**, 22–29.

Klite, P. D. and Diercks, F. H. (1965). *Histoplasma capsulatum* in fecal contents and organs of bats in the canal zone. *Am. J. trop. Med. Hyg.* **14**, 133–139.

Koch, H. A. (1965). Zur Ökologie von *Trichophyton verrucosum* Bodin 1902. *Proc. Int. Symp. Med. Mycol. Warsaw*, 1963, 75–79.

Lutz, A. and Splendore, A. (1907). Uber eine bei Menschen und Ratten beobachtete Mykose. Ein Beitrag zur Kenntniss der sogenannten Sporotrichosen. *Zentbl. Bakt.ParasitKde*, (Orig.) **45**, 631–637.

McDiarmid, A. (1955). Aspergillosis in free-living wild birds. *J. comp. Path. Ther.* **65**, 246–249.

McDiarmid, A. (1962). Diseases of free-living wild animals. *F.A.O. agr. Stud.* No. 57, 1–119.

McDiarmid, A. and Austwick, P. K. C. (1954). Occurrence of *Haplosporangium parvum* in the lungs of the mole (*Talpa europaea*). *Nature, Lond.* **174**, 843–844.

Mackenzie, P. Z. and Simpson, R. M. (1956). The African Veterinary Handbook. 2nd edit. *Pitman, Nairobi*, 281.

McKeever, S., Kaplan, W. and Ajello, L. (1958a). Ringworm fungi of large wild mammals in Southwestern Georgia and Northwestern Florida. *Am. J. vet. Res.* **19**, 973–975.

McKeever, S., Menges, R. W., Kaplan, W. and Ajello, L. (1958b). Ringworm fungi of feral rodents in Southwestern Georgia. *Am. J. vet. Res.* **19**, 969–972.

Maddy, K. T. (1960). Coccidioidomycosis. *Adv. vet. Sci.* **6**, 251–256.

Mariat, F. and Segretain, G. (1956). Étude mycologique d'une Histoplasmose spontanée du singe africain (*Cynocephalus babuin*). *Annls Inst. Pasteur, Paris* **91**, 874–891.

Medical Research Council (1967). Nomenclature of fungi pathogenic to man and animals. *Med. Res. Coun. Memo No.* 23. 3rd. edit. 20 pp.

Menges, R. W., Love, G. J., Smith, W. W. and Georg, L. K. (1957). Ringworm in wild animals in Southwestern Georgia. *Am. J. vet. Res.* **18**, 672–677.

Montague, G. (1813). Supplement to the Ornithological Dictionary (1802). S. Woolmer, London.

Neff, J. A. (1955). Outbreak of aspergillosis in mallards. *J. Wildl. Mgmt* **19**, 415–416.

Nicolet, J., Klinger, K., and Fey, H. (1967). *Dermatophilus congolensis* agent de la streptotrichose du chamois. *Path. Microbiol.* **30**, 831–837.

Otčenášek, M. and Dvořák, J. (1962). The isolation of *Trichophyton terrestre* and other keratinophilic fungi from small mammals of South Eastern Moravia. *Sabouraudia* **2**, 111–113.

Owen, R. (1832). Notes on the anatomy of the flamingo. (*Phaenicopteris ruber* L.). *Proc. zool. Soc. Lond.* **1832**, 141–144.

Paldrok, H. and Zetterberg, B. (1962). A contribution to the study on the occurrence of adiaspiromycosis (haplomycosis) in rodents in Sweden. *Acta path. Microbiol. scand.* **56**, 65–69.

Pätiälä, R. (1951). On fungus diseases in game. *Riistat. Julk.* **6**, 21–22.

Poulding, R. H. (1957). Tuberculosis in gulls: a preliminary investigation. *Bull. Br. Orn. Club* **77**, 144–149.

Pringault, E. and Vigne, P. (1921). Note sur un cas de sporotrichose naturelle du rat. *Bull. Soc. fr. Derm. Syph.* **28**, 342–344.

Prokopič, J., Dvořák, J., Otčenášek, M. (1965). Adiaspiromycosis (Haplomycosis) in Czechoslovakia. Preliminary report. *Sabouraudia* **4**, 35–36.

Rousseau, E. and Serrurier (1841). Développement de cryptogames sur les tissues de vertébrés vivants. *C. T. hebd. Séanc. Acad. Sci., Paris* **13**, 18–19.

Smith, J. M. B. and Marples, M. J. (1963). *Trichophyton mentagrophytes* var. *erinacei*. *Sabouraudia* **3**, 1–10.

Smith, J. M. B. and Austwick, P. K. C. (1967). Fungal diseases of rats and mice. *In* "Pathology of Laboratory Rats and Mice". pp. 681–732, (E. C. Cotchin and F. J. C. Roe, eds.), Blackwell, Oxford.

Smith, W. W., Menges, R. W. and Georg, L. K. (1957). Ecology of ringworm fungi on commensal rats from rural premises in Southwestern Georgia. *Am. J. trop. Med. Hyg.* **6**, 81–85.

Staib, F. (1962). Vogelkot ein Nahrsubstrat fur die Gattung *Cryptococus. Zentbl. Bakt. ParasitKde* (Abt. 1) **186**, 233–247.

Swatek, F. E. and Plunkett, O. A. (1957). Ecological studies on *Coccidioides immitis. Proc. Symp. Coccidioidomycosis U.S. Dept. Health, Educ. and Welfare.* 161–167.

Taylor, W. W., Radcliffe, F. and van Pleenen, P. F. D. (1964). A survey of small Egyptian mammals for pathogenic fungi. *Sabouraudia* **3**, 140–142.

Vanbreuseghem, R. (1963). Influence of fungi on wildlife population. *Proc. 1st Int. Congr. Wildl. Dis. New York* 1962. 96–136.

Van Uden, N. (1963). Factors of host-yeast relationship. *Recent Prog. Microbiol.* **8**, 635–643.

Walker, J. and Spooner, E. T. C. (1960). Natural infection of the African Baboon *Papio papio* with the large cell form of *Histoplasma. J. Path. Bact.* **80**, 436–438.

DISCUSSION

COOPER: Were the traps Dr. Blackmore and Miss Owen used, "breakback" or live traps? If the former, how often were they re-examined in order to ensure that no parasites had escaped?

OWEN: We use the live traps in preference to the others for, in fact, the reasons you describe.

MOLYNEUX: I do not think *Trypanosoma lewisi* is pathogenic for rats as, I think, was suggested by Dr. Blackmore, neither are the trypanosomes of voles; the trypanosomes of voles can be transmitted to many species of fleas.

HERMAN: It was pointed out this morning that stress is an important factor in wildlife disease and I think we should emphasize that this has been brought about by habitat relationships that are associated with many diseases. For example, we find aspergillosis very common every Spring when the food supply of migrating waterfowl is frozen up for a few days and the birds are weakened from their northward migration; this includes swans, geese and ducks in the middle of this migration and we get many cases every year. These birds are found sick within a week or two after the freeze-up and they are usually heavily infected with aspergillus.

For about ten years we have been studying the cause of death in wintering canadian geese that take refuge in the southern range. This is a sand island a mile wide and ten miles long and 5000 to 10 000 canadian geese are present here in the wintertime. The normal food consists of roots of a particular grass growing in the body of water that separates this island habitat from the mainland. The water is normally six to twelve inches deep; however, when storms occur the water may pile up to three feet deep and the birds cannot get at the roots. Even although there is ample food ten miles away across the water, they prefer for some reason to remain on this island and in two or three days the limited food on land is consumed and malnutrition appears in the birds. Moreover, infections with gizzard worms becomes serious.

AUSTWICK: I would like to ask Dr. Herman how long he thinks it takes for a chronic case of aspergillosis to develop in wildfowl. I am puzzled at the moment as to whether it might be a week or three months.

HERMAN: In the laboratory we have developed a technique of transmitting aspergillosis by inoculating spores into the airsac; in such cases by using baby ducks or chicks we can produce severe airsac involvement and typical tubercle-like lesions in the lung. The birds are either dead in ten days or recover. If we look at them several weeks later, the lesions are all gone. Evidence from the wild indicates waterfowl may survive for several weeks and still show extensive lesions.

BRAMLEY: I would like to make another observation about what Dr. Dunn was saying concerning liver fluke and lung worm and this perhaps backs up the importance of stress and social status of the animals involved. I am engaged in an ecological study of roe deer and it is my experience that the animals which go down most frequently with liver fluke and lung worm are the younger male animals ejected out of the population in the woods. These are the ones undoubtedly under stress whereas the established males in their own particular territories seem to be reasonably resistant to these infections. I do not know whether Dr. McDiarmid's observations are similar to mine or not.

DUNN: We have not done much of this kind of observation in this country but I think it is true to say that some of the comparative work in America, studying, for instance, deer dying off in the winter and deer killed by shooting, has shown heavier parasitic infections in winter-killed deer. Although there is no doubt that, in nature, animals under stress tend to carry heavier infections, it is still impossible to substantiate the connection in terms of pathogenesis and resistance.

MURTON: The meeting seems to be showing much interest in the extent to which animals suffering from so-called stress syndromes may be predisposed to overt disease or the development of subclinical infections. I should like to make two points arising from my studies of the woodpigeon (*Columba palumbus*).

First, woodpigeons suffering from avian tuberculosis are more easy to shoot than healthy birds. Of 605 shot between January and March 1959/61, 3·1% were diseased, whereas of a comparable sample of 122 birds collected with stupefying baits, which are seemingly unbiased in this respect, only 1·6% were affected. Clearly, the method of obtaining field samples needs definition when discussing disease incidence in wild subjects.

Second, woodpigeons have a very marked social hierarchy which is manifested in the flock structure (see *J. appl. Ecol.* **3**, 55–96 (1966). Birds which are low in the social hierarchy spend a disproportionate amount of their time avoiding intraspecific conflict situations and as a result obtain insufficient food. When such individuals are collected from the flocks, they often prove to be emaciated and to have considerably enlarged adrenal glands compared with their flock mates. Furthermore, they more often show symptoms of tuberculosis than the other individuals and, with the samples examined, the differences are statistically significant. It seems reasonable to suppose that overt manifestations of tuberculosis would not have developed had these birds not been already low in the social heirarchy. It should also be mentioned that these individuals are more easy to shoot than healthy birds and that it is "stressed" birds which are featuring more often in the shot samples of unknown origin. I would attach only secondary importance to disease as a factor regulating pigeon numbers using arguments similar to those adopted in the case of endocrine stress (see *Ibis* **109**, 622–623 (1967).

I would like to ask one last question; when these woodpigeons are being looked at, what is the incidence rate of tuberculosis in the two groups. As far as we are concerned, the pigeons shot vary according to the time of year. We have had a range from 4·0–6·0% but maybe, as you say, it is a question of the easier shots.

COOPER: On the subject of aspergillosis, in my opinion, stress plays a part here. I have done quite a few post mortem examinations on imported birds of prey and aspergillosis is common in these. In fact, they often die a week to ten days after arrival, even when transported singly. These birds undergo particular stress in capture and transport and are frequently in a poor nutritional state upon arrival.

CHAIRMAN'S SUMMING UP

In summing up, I should like to mention one specific point concerning the term "stress". There appears to be two interpretations of "stress": (i) to describe a specific pathological situation and (ii) to describe a subjective assessment by a human observer. It would seem to me that the main duty of the person who collects and collates our discussions would be to distinguish between these two interpretations when they have arisen in our discussions this afternoon.

Next I would like to thank the speakers for giving us such entertaining and informative talks. I am sure you will agree that they are doing extremely interesting and systematic work designed to find out more and more about the various situations in the wildlife populations, situations which may well have an effect on the health of our domesticated animals and of human beings.

Finally, I should like to thank all the contributors to the discussion.

NEOPLASMS

CHAIRMAN: E. COTCHIN

CHAIRMAN'S INTRODUCTION

There are two more papers for discussion, both included under the heading of "Neoplasms". This is perhaps a little misleading. The paper on myxomatosis appears at this part of the session because the contribution by Dr. Jennings eventually proved to be somewhat shorter than anticipated and, also, the virus part of the Conference was initially oversubscribed. However, it does raise a neat little point, namely, is myxomatosis in fact a neoplasm? Before Dr. Jennings' contribution I would like to make one or two general comments about neoplasms in wild animals. It is perhaps difficult nowadays to conceive that for a long time it was thought that neoplasia, including cancer, was a disease strictly confined to humans, even perhaps a disease of civilization. In the first decade of the last century, 1802 to be precise, the Edinburgh Medical Society, in some farsighted published questions which were designed to stimulate an enquiry into human cancer, raised the question as to whether anything like cancer did affect the brute creation and it is also of interest that the earliest programme of the Imperial Cancer Research Fund included a study of the spontaneous occurrence of tumours in animals. Fortunately, that is from the human point of view, it has turned out that cancer is by no means confined to man and our understanding of the human disease would have been incomparably the poorer if this had not been the case. A good deal of information is now available about tumours of domestic and laboratory animals as well as zoo animals. In the most general terms, what has emerged from these studies is the opportunity for the investigation of the aetiology, biological behaviour and therapy of cancer that would otherwise have been impossible or, at least, very difficult to accomplish. In this connexion it is only necessary to refer to such well known virus-induced tumours as the Rous sarcoma of fowls, avian leukosis, mammary cancer and leukaemia in mice. Two particular aspects, worth further mention, are the occurrence of a more or less characteristic spectrum of types of tumour in each of the different species so far studied in any depth and the widespread occurrence of malignancy of the lymphoid tissue (that is,

273

lymphosarcoma) in a whole range of species; this tumour has, of course, now shown evidence of virus aetiology in domesticated as well as laboratory animals. The study of tumours in wild animals, that is, in addition to zoo animals and other species, seems to me to be well worth pursuing. Already some very interesting information has been gained from them—I might mention, as an example, the viral induced renal carcinoma of the leopard frog in America. The search for other tumours which can be studied aetiologically and in other ways is very desirable. I am sure Dr. Jennings' paper will be a valuable contribution to a field of knowledge which still needs much more extensive cultivation.

Symp. zool. Soc. Lond. (1968) No. 24, 273–287.

TUMOURS OF FREE-LIVING WILD MAMMALS AND BIRDS IN GREAT BRITAIN

A. R. JENNINGS

Department of Animal Pathology, University of Cambridge, England

SYNOPSIS

Spontaneous tumours of free-living wild mammals and birds are apparently rare not only in Britain but throughout the world. Existing records are sparse and widely scattered throughout the literature in journals covering very different disciplines. In contrast, neoplasia in wild species in captivity is not uncommon and is well documented.

This paper summarizes existing information on neoplastic conditions seen in wild-life in Britain and analyses the results. A total of forty-four avian and eleven mammalian tumours have been tabulated; these occurred in twenty-one species of birds representing thirteen families and in five species of mammals. Several tumours are selected for comment, in particular the papilloma of the skin of the chaffinch, the subcutaneous fibroma and the lymphosarcoma of deer. Comparisons are made with some aspects of neoplasia in captive wildlife.

This contribution serves to illustrate the lack of detailed information on the subject; even where cases are recorded the information is usually incomplete. Much useful information has probably already been lost and will continue to be lost unless accurate and detailed records are kept. It is likely that advances in knowledge on neoplasia in man and domestic stock could be made by extending the scope of comparative oncology to include free-living species of wildlife.

INTRODUCTION

Much of the information on the infectious diseases of free-living wild-life has been brought together by McDiarmid (1962) but the literature on neoplasia in wildlife is widely scattered in journals representing many scientific disciplines. The bibliography is sparse and the cases recorded, in general, are not well documented. On the other hand, there is a good deal of information available on neoplastic disease in captive wild mammals and birds. The excellent and authoritative bibliography of references to disease in wild mammals and birds, mainly those in captivity, by Halloran (1955) covered the literature up to that point in time and an extension of that survey is now needed. It is certain that much useful information is being lost due to failure to record cases, inefficient recording and/or failure to co-ordinate existing information. The advantages to be gained from comparative pathology are already apparent in many fields and it is certain that comparative oncology would benefit considerably by including neoplasia of wildlife.

During the last decade there have been three main surveys into the causes of death in free-living wild birds in Britain (Keymer, 1958; Jennings, 1961 and Macdonald, 1965a). As a result of these surveys much useful information has resulted from the autopsy examinations of several thousand birds. Unfortunately I know of no similar surveys being made into mammalian mortality although there have been surveys into mortality of particular species, e.g. foxes (Blackmore, 1964). This failure to make use of potentially valuable pathological material is to be regretted especially when one considers the very large numbers of wild animals killed annually in Britain.

The tumours discussed in this paper have all occurred in British free-living wild birds and mammals. They do not include cases from animals in semi-captivity such as those in parks and zoological gardens, nor are semi-feral species included; myxomatosis is not discussed because this will form a separate paper in this Symposium. The avian tumours investigated were noted during the surveys into bird mortality previously mentioned, some of these tumours were recorded elsewhere and several are previously unrecorded cases sent to me by various research workers to whom acknowledgements are made.

RESULTS

A total of 44 avian and 11 mammalian cases of neoplasia was recorded. The details of the species involved, the type of tumour and the sites of origin, together with the appropriate reference, are given in Table I. The avian tumours occurred in 21 species representing 13 families and 5 species of mammals. Twenty two or 50% of the avian tumours could be classified as malignant and 5 of the 11 mammalian ones were malignant, using as criteria of malignancy those suggested by Willis (1960). The common benign tumour of birds in this survey was the cutaneous papilloma and in deer there were equal numbers of cases of skin papillomata and fibromata. The lymphosarcoma apparently was the common malignant neoplasm of fallow and roe deer in Britain.

DISCUSSION

No attempt has been made to assess the incidence of neoplasia because the total number of animals and birds examined is not known. However, a tentative figure can be arrived at from some of the surveys where numbers are known accurately and it would appear that in wild birds the incidence is probably between 0·1 and 1%. In domestic poultry there is a much higher incidence, the figure may be as high as

27%, according to Feldman and Olson (1959), but in captive wild birds the incidence again is low, e.g. Lombard and White (1959) in Philadelphia demonstrated an incidence of 1·4% in 10 240 autopsies. It is generally suggested that the low incidence is in part due to the relatively short life span and in support of this Lombard and White (1959) noted that the tumour incidence was roughly parallel with the period of time the birds were on exhibition. On the other hand, an interesting observation was put forward by Snyder and Ratcliffe (1966) who stated that during the period of 1943–1961 the annual longevity in one collection of birds had not changed and the increased frequency of neoplasia could not be explained on these grounds. They speculated that an increase in atmospheric carcinogens was a possible causal factor in the increased incidence. Although it seems probable that in the wild state the relatively short life span of many species is an important factor in determining the incidence of malignant neoplasia, this cannot be the complete answer. In the domestic fowl the life span is becoming progressively shorter and yet the incidence of neoplasia remains high. It is probable that species susceptibility plays a part in determining tumour incidence.

Blackmore (1965) and others have also drawn attention to avian species variation and Blackmore emphasized the unique susceptibility of the budgerigar (*Melopsittacus undulatus*) to neoplastic disease. The same author also pointed out that there appeared to be almost no studies of disease in free-living Australian budgerigars. Although the figures in this survey are scanty can it be significant that tumours have so far only been recorded in some 21 species out of a total of some 550 birds on the British list? A total of 11 cases occurred in waterfowl or in sea birds or in birds associated with water.

The chief organs involved in these naturally occurring cases of avian neoplasia are the skin, the urogenital system, the liver and the musculo-skeletal system. If we look at domestic fowls we see that skin tumours are not common and they are also infrequent in captive wild birds. The high incidence of skin tumours in this series is due entirely to papillomatosis in chaffinches. Out of a total of 44 avian tumours, 16 were papillomata and all these cases occurred in chaffinches (*Fringilla coelebs*). These chaffinch tumours have been described by various authors in Britain, in Germany and in Sweden, e.g. Macdonald (1965b). The neoplasm usually affects the claws and may be uni- or bi-lateral. Much less often affected is the skin at the junction of the beak and the face. The tumours are greyish brown in colour with a rough irregular surface and a typical papillomatous appearance. Histologically they are benign neoplasms. They may attain a large size and extend to a diameter of 2 cm, such large tumours may cause mechanical disability so that the

TABLE I

Neoplasia in free-living wild mammals and birds in Britain

Species	Tumour and site	Reference
Chaffinch (*Fringilla coelebs*) (2 cases)	Papilloma. Claws	Jennings (1959)
Chaffinch (*Fringilla coelebs*) (3 cases)	Papilloma. Claws	J. W. Macdonald (1968, personal communication)
Chaffinch (*Fringilla coelebs*) (4 cases)	Papilloma. Claws	Macdonald (1965b)
Chaffinch (*Fringilla coelebs*)	Not diagnosed. Skin, legs	J. W. Macdonald (1968, personal communication)
Chaffinch (*Fringilla coelebs*) (6 cases)	Papilloma. Legs and toes	Keymer and Blackmore (1964)
Chaffinch (*Fringilla coelebs*)	Papilloma. Skin	J. W. Macdonald (1965b)
House sparrow (*Passer domesticus*)	Clear cell carcinoma. Kidney	J. W. Macdonald (1968, personal communication)
Oystercatcher (*Haematopus ostralegus*)	Seminoma. Testis	J. W. Macdonald (1968, personal communication)
Oystercatcher (*Haematopus ostralegus*)	Osteo-fibroma-lipoma. Mesentery	J. W. Macdonald (1968, personal communication)
Lapwing (*Vanellus vanellus*)	Schwann cell tumour. Rt. brachial plexus	J. W. Macdonald (1968, personal communication)
Red grouse (*Lagopus scoticus*)	Plasmacytoma. Eyelids	J. W. Macdonald (1968, personal communication)
Red grouse (*Lagopus scoticus*)	Multiple fibromata. Skin	G. W. Storey (1968, personal communication)
Mallard (*Anas platyrhynchos*)	Dysgerminoma. Ovary	G. W. Storey (1968, personal communication)
Mallard (*Anas platyrhynchos*)	Granulosa cell. Ovary	Beer and Storey (1961)
Greenfinch (*Chloris chloris*)	Adenocarcinoma. Liver	Jennings and Soulsby (1957)
Black-headed gull (*Larus ridibundus*)	Cystadenoma. Ovary	Jennings and Soulsby (1957)
Blue tit (*Parus caeruleus*)	Sarcoma. Intestine	Jennings and Soulsby (1956)
Wood pigeon (*Columba palumbus*)	Spindle cell sarcoma. Sternum	Jennings and Soulsby (1956)
Wood pigeon (*Columba palumbus*)	"Leucosis-like." Viscera	Jennings (1955)

Species	Tumour	Reference
[Wo]od Pigeon (Columba palumbus)	Pleomorphic myosarcoma. Wing muscles	G. W. Storey (1968, personal communication)
Kittiwake (Rissa tridactyla)	"Leucosis-like." Liver	Jennings (1955)
Shelduck (Tadorna tadorna)	"Leucosis-like." Viscera	Jennings (1954)
Shelduck (Tadorna tadorna)	Lymphofollicular reticulosis. Liver; spleen	G. W. Storey (1968, personal communication)
Partridge (Perdix perdix)	Neural lymphomatosis. Sciatic nerve	Jennings (1954)
Little owl (Athene noctua)	"Leucosis-like". Viscera	Jennings (1954)
Little owl (Athene noctua)	Adenocarcinoma. Liver	Blackmore (1965)
Blackbird (Turdus merula)	Adenocarcinoma. Kidney	Jennings (1959)
Pheasant (Phasianus colchicus)	Arrhenoblastoma. Ovary	G. W. Storey (1968, personal communication)
Wigeon (Anas penelope)	Fibrosarcoma. Lung muscles	G. W. Storey (1968, personal communication)
Robin (Erithacus rubecula)	Haemangioma. Eyelid	G. W. Storey (1968, personal communication)
Pochard (Aythya ferina)	Adenocarcinoma. Lung, liver	G. W. Storey (1968, personal communication)
Linnet (Carduelis cannabina)	Fibroma. Sternum	A. R. Jennings (1968, personal communication)
Red deer (Cervus elaphus) (2 cases)	Multiple Fibromata. Skin	A. McDiarmid (1968, personal communication)
Red deer (Cervus elaphus) (2 cases)	Multiple Papillomata. Skin, abdomen	A. R. Jennings (1968, personal communication)
Fallow deer (Dama dama)	Lymphosarcoma.	A. McDiarmid (1968, personal communication)
Fallow deer (Dama dama)	Lymphosarcoma.	A. McDiarmid (1968, personal communication)
Fallow deer (Dama dama)	Haemangiosarcoma. Liver	A. McDiarmid (1968, personal communication)
Roe deer (Capreolus capreolus)	Lymphosarcoma.	McDiarmid (1967)
Roe deer (Capreolus capreolus)	Lymphosarcoma. Viscera	Woodford (1966)
Hedgehog (Erinaceus europaeus)	Adenoma. Pituitary	Campbell and Smith (1966)
Coypu (Myocastor coypu)	Adenoma. Kidney	A. Levene (1968, personal communication)

tarsus is flattened from side to side and there may be loss of one or more claws. There appears to be no sex or age incidence but a number of cases may occur at the same time, suggesting a common cause. It seems not unlikely that the papilloma of the claws of chaffinches is an infectious neoplasm and is worthy of further investigation. Why it should be confined to this particular species and not apparently involve other finches is curious, particularly when it is recalled that this finch is gregarious and outside the breeding season is usually in company with large flocks of greenfinches, bramblings and so forth. Yet the claw papilloma has not been recorded from these other species. On the other hand, avian pox produces a somewhat similar gross lesion; it also affects small passerines and may cause confusion in diagnosis unless histological examinations are carried out. The pox virus is almost certainly spread by contamination of perching sites. Is this also the case with the papilloma of the chaffinch?

In this series of tumours of free-living wild birds there were 7 cases involving the urogenital system, i.e. 14%, and of these 5 involved the ovary. The figures are small for anything other than conjecture but it may be significant that two ovarian tumours were in mallard. Ovarian tumours are not infrequent in domestic poultry and in captive wild birds. It has long been recognized that the psittacine species are especially liable to disease of the urogenital system and especially to tumours of the gonads but no reasons for this are apparent. Not infrequently the gonadal tumours of the budgerigar are functional and bring about sex reversal and similar observations have been made in the domestic fowl although such cases are rare. An arrhenoblastoma was seen in a pheasant (G. W. Storey, 1968, personal communication) and the same author described a granulosa cell tumour and a dysgerminoma both in mallard. J. W. Macdonald (1968, personal cummunication) has recorded a seminoma in an oystercatcher.

Tumours involving the respiratory system were few in number in this series. On the other hand, Stewart (1966), discussing tumours of captive wild mammals and birds seen at post mortem examination, commented on the relatively high incidence of lung tumours. The chief type in his series was the adenocarcinoma and over 50% of the pulmonary cancers occurred in the family *Anatidae*. Snyder and Ratcliffe (1966) also commented on this curious organ susceptibility of the *Anatidae* and within this family singled out for special comment the shoveler (*Spatula clypeata*) and the red head (*Oxyura jamaicensis*). Despite the paucity of material is it perhaps significant that in this survey a fibrosarcoma of the lung muscles was seen in a wigeon (*Anas penelope*) and an adenocarcinoma of the lung in a pochard (*Aythya ferina*)?

It is an interesting fact that in domestic poultry of all tumours those of the lymphoid and haemopoietic tissues are the most common but no comparable unequivocal cases have yet been encountered in free-living wild birds. However, three cases of a condition morphologically indistinguishable from Marek's disease have been seen with lesions in the viscera in two cases and with neural lesions in a third. The first two cases were in a kittiwake (*Rissa tridactyla*) and a little owl (*Athene noctua*) and the third in a partridge (*Perdix perdix*). Is it possible that wild birds play a part in the epidemiology of this disease of the domestic fowl? It is easy to see how the partridge could become infected but less easy to understand infection of the other two species. However, according to Biggs and Payne (1967), Marek's disease is highly contagious and can be spread by direct and indirect contact. They cite references indicating that the litter may be infective for up to six weeks after contact with infected birds and that the agent can survive for long periods in the darkling beetle (*Alphitobius diaperinus*). It is possible that infection of the little owl might have arisen in such a way as this.

Turning now to the mammalian tumours, the largest number has been seen in deer. The multiple papillomata recorded were seen on the skin of the red deer on the Island of Rhum (A. R. Jennings, 1968, personal communication) and they occurred chiefly on the ventral abdomen and on the latero-ventral aspects of the neck. They resembled the papillomata seen in domestic stock and require no further comment except to point out that this tumour is quite distinct from the lesion called the fibropapilloma. This neoplasm is best regarded as a fibroma and I am grateful to Dr. A. McDiarmid (1968, personal communication) for providing examples of this lesion. These tumours, taken from red deer (*Cervus elaphus*) in Scotland, appear to be identical on macroscopic and histological grounds with the tumours investigated by Shope, Mangold, MacNamara and Dumbell (1958). These authors described a naturally occurring cutaneous fibroma of the Virginia white-tailed deer (*Odocoileus virginianus*) which was of viral origin being readily filterable and capable of surviving for at least 27 months in glycerol saline at $-20°C$. The neoplasm could be transmitted to other deer and had an incubation period of about seven weeks; the tumour had a high regression rate. Fay (1962) reviewed the existing literature on the subject and recorded 9 deer in Michigan with similar skin tumours.

These tumours, both British and American, are firm and white on section. Histologically they are made up of many stellate, angulated and spindle-shaped cells which resemble fibroblasts at different stages of development. There is variable collagen formation. The collagen is present in whorls; in some areas it is dense but in others the amount of collagen is minimal. Together with this there is also myxomatous tissue and quantities of an acid muco-polysaccharide ground substance. The overlying epithelium is greatly thickened and heavily pigmented. This thickened and cornified epithelial layer gives the tumour the superficial appearance of a papilloma and the tumour is sometimes described as a fibro-papilloma. The lesion bears a close resemblance to the rabbit and squirrel fibromata but no acidophilic inclusions could be demonstrated in the cases available. This, however, does not necessarily imply that they are absent since, as pointed out (Jennings, 1966), viral inclusions in general are not always present throughout the disease process. Their staining characteristics are variable and much depends upon methods of fixation and staining. It is certain that further work with the deer fibroma would be of great interest. McDiarmid (1967) posed the problem as to whether cattle could be infected with this neoplasm or conversely could cattle infect deer with it. On the other hand, Shope *et al.* (1958) failed to demonstrate any relationship between the deer neoplasm and the bovine papilloma. Has the deer neoplasm the same curious characters of the squirrel fibroma and is there any relationship between these tumours?

Of all tumours, lymphatic leukaemia and the lymphosarcoma have probably the widest range of occurrence in domestic stock and in some species they may well be the most important tumours. Lymphosarcoma occurs in fallow (*Dama dama*) and roe deer (*Capreolus capreolus*) and probably in red deer (*Cervus elaphus*), also. In cattle there is an ever growing weight of evidence to indicate that neoplasia of lymphoid tissue is of viral origin and a like aetiology has already been demonstrated for the cat (Jarrett, Martin, Crighton, Dalton and Stewart, 1964), similar evidence is impressive in the dog (Owen and Nielsen, 1968). It seems likely that the lymphosarcoma of deer may also have a similar cause. In cattle there is evidence of horizontal transmission of the disease by movement of animals from infected to non-infected herds and also inadvertently by infected blood during premunity vaccination with piroplasms. With deer in the wild state, movement from herd to herd is not likely because of territory habits but tick-borne infection is not at all unlikely.

The morphology of the lesions and their distribution in deer closely resemble the bovine disease. The cytology of the tumour varied somewhat in the cases seen. Generally, there is widespread invasion of the affected organ with loss of the normal architecture. The cells are irregularly spherical with round nuclei which are strongly basophilic. There is an abundance of small granular chromatin and mitotic figures in some areas are prominent.

The age incidence of bovine lymphosarcoma is of interest (Cotchin, 1966) and even from the few cases in deer it appears as if there may be a further similarity because both the thymic form (6–30 months) and the adult form (5–8 years) have been seen in the few cases available.

Lymphatic leucosis in cattle may occur in two phases with a good deal of overlap between them. There is a phase of lymphocytosis and a subsequent phase with extensive involvement of the lymph nodes and the viscera. So far as I am aware, no observations have ever been made on the blood picture of deer with lymphatic leucosis and yet much is known about the blood picture of deer especially in relation to the phenomenon of sickling. A re-examination of old blood films may be well worthwhile and future attention to this point when examining deer culled for various reasons would, I am sure, be a useful exercise.

The only other malignant neoplasm seen in wild deer in Britain was in a three-year-old fallow buck (*Dama dama*) and it is of interest in that the malignant neoplasm was confined to the liver (A. McDiarmid, 1968, personal communication). The liver was greatly enlarged and weighed $18\frac{1}{2}$ lb. it had a nutmeg appearance on section. There were many variously sized swellings, some clearly demarcated, others haemorrhagic and necrotic. The histology is greatly complicated with extensive thrombosis of the portal veins which has produced large areas of necrosis. The neoplastic cells are poorly differentiated but appear to be of mesenchyme origin. It is sarcomatous, probably a fibrosarcoma, but there are areas of new blood vessels which might indicate that the lesion is a haemangiosarcoma.

Turning now to other species of mammals, Campbell and Smith (1966) recorded an interesting case of a pituitary adenoma in a hedgehog (*Erinaceus europaeus*). The lesion was a spontaneous chromphobe tumour which caused partial obstruction to the flow of the neuro-secretion. The tumour was on the left lobe of the pars distalis but did not cause clinical affects although the appearance of the tumour suggested that it had been present during at least one period of hiber-nation. It is believed that hibernation is associated with a temporary regression of the endocrine system and it has been shown that experi-mental tumours will fail to grow during hibernation (there is an inter-esting parallel here with the failure of viruses to multiply and cause disease in hibernating animals, e.g. foot-and-mouth virus). When the animals emerge from hibernation both tumour cells and viruses start to multiply. It is certain that an investigation into the mechanisms of multiplication inhibition would be very profitable both from the point of oncogenesis and of virus replication. Campbell and Smith (1966) point out that another significant point about this hedgehog tumour is the fact that the evaluation of experimental evidence may be wrongly assessed because of unsuspected disease in the experimental animal.

It is somewhat surprising that spontaneous tumours have not been recorded in wild rats and mice in Britain in view of the high incidence in laboratory stock and of the fact that large numbers of rats have been examined for evidence of such diseases as salmonellosis. An interesting paper relevant to this is that by Andervont and Dunn (1962). These workers described the occurrence of spontaneous tumours in wild house mice (*Mus musculus*) raised in captivity. Forty-three % of these mice showed tumours on necropsy. Most tumours were found in the older age groups and there was a relatively high incidence of pulmonary tumours and of reticulum cell tumours. Another curious feature is that viruses responsible for leukaemia in laboratory strains of mice are well documented but wild mice apparently rarely developed lymphocytic tumours. If this observation is correct then is it that wild mouse strains are more resistant to lymphocytic leukaemia than most inbred strains? Andervont and Dunn also pose the question as to whether wild strains carry a virus capable of producing leukaemia in inbred strains. They also discuss the occurrence of spontaneous hepatomas in wild mice and indicate that there is some evidence that wild males are more susceptible to this tumour than laboratory mice. In view of these findings in captive wild mice, it would, I feel, be well worthwhile investigating the incidence of spontaneous neoplasia in free-living rodents. Such an investigation would be relatively simple.

So far as I am aware, to date only one tumour has been recorded in a free-living rodent in Britain. This was a papillary adenoma of the kidney of a coypu (*Myocastor coypus*) (A. Levene, 1968, personal communication). The animal was a 1–2-year-old male with multiple buff coloured nodules in one half of a kidney.

SUMMARY

A total of 44 avian and 11 mammalian tumours has been investigated. The neoplasms occurred in 21 species of birds and in 5 species of mammals. Tumours worthy of special comment are the fibromata and the lymphosarcomata of deer and the skin papillomata of chaffinches.

ACKNOWLEDGEMENTS

The following have contributed material and/or information which has been essential in the preparation of this paper and I am most grateful for their assistance: Drs J. V. Beer, D. K. Blackmore, Professor E. Cotchin, Drs I. F. Keymer, A. Levene, J. W. Macdonald, A. McDiarmid and G. W. Storey.

References

Andervont, H. B. and Dunn, T. B. (1962). Occurrence of tumours in wild house mice. *J. nat. Cancer Inst.* **28**, 1153–1163.

Beer, J. V. and Storey, G. W. (1961). An ovarian tumour in a mallard, *Anas platyrhynchos*. *Bull. Br. Orn. Club* **81**, 153–156.

Biggs, P. M. and Payne, L. M. (1967). The avian leucosis complex. *Vet. Rec.* **80**, v–vii, Clin. Supplement 22 April.

Blackmore, D. K. (1964). A survey of disease in British wild foxes, *Vulpes vulpes*. *Vet. Rec.* **76**, 527–533.

Blackmore, D. K. (1965). The pattern of disease in budgerigars: a study in Comparative Pathology. Ph.D. Thesis. London, 1965.

Campbell D. J. and Smith, W. T. (1966). A pituitary adenoma in a hedgehog, *Erinaceus europaeus*. *Endocrinology*. **79**, 842–844.

Cotchin, E. (1966). Some aetiological aspects of tumours in domesticated animals. *Ann. R. Coll. Surg.* **38**, 92–116.

Fay, L. D. (1962). Neoplastic diseases of white-tailed deer. *Proc. natn. White-tailed deer Symp.* No. 1. 132–137.

Feldman, W. H. and Olson, C. (1959). Neoplastic diseases of chickens. *In* "Diseases of Poultry". (H. E. Biester and L. H. Schwarte, eds.) pp. 642–700, Iowa State College Press, Ames, Iowa, U.S.A.

Halloran, P. O'C. (1955). A bibliography of references to diseases in wild mammals and birds. *Am. J. vet. Res.* **16**, Supplement.

Jarrett, W. F. H., Martin, W. B., Crighton, G. W., Dalton, R. G. and Stewart, M. F. (1964). Transmission experiments with leukaemia (lymphosarcoma). *Nature, Lond.* **202**, 566–567.

Jennings, A. R. (1954). Diseases in wild birds. *J. comp. Path.* **64**, 356–359.

Jennings, A. R. (1955). Diseases in wild birds. *Bird Study* **2**, 69–72.

Jennings, A. R. (1959). Diseases in wild birds, Fifth Report. *Bird Study*, **6**, 19–22.

Jennings, A. R. (1961). An analysis of 1000 deaths in wild birds. *Bird Study* **8**, 25–31.

Jennings, A. R. (1966). *In* "The Histopathological Diagnosis of Viral Disease". (W. A. Pool, ed.), *Vet. A.* 7th issue. pp. 98–106. Bristol: John Wright and Sons.

Jennings, A. R. and Soulsby, E. J. L. (1956). Diseases of wild birds, Third Report. *Bird Study* **3**, 270–272.

Jennings, A. R. and Soulsby, E. J. L. (1957). Diseases of wild birds, Fourth Report. *Bird Study* **4**, 216–220.

Keymer, I. F. (1958). A survey and review of the causes of mortality in British birds and the significance of wild birds as disseminators of disease. *Vet. Rec.* **70**, 713–720; 736–740.

Keymer, I. F. and Blackmore, D. K. (1964). Diseases of the skin and soft parts of wild birds. *Br. Birds* **57**, 175–179.

Lombard, L. S. and White, E. J. (1959). Frequency and types of tumours in mammals and birds of the Philadelphia Zoological Garden. *Cancer Res.* **19**, 127–141.

Macdonald, J. W. (1965a). Mortality in wild birds. *Bird Study* **12**, 181–195.

Macdonald, J. W. (1965b). Chaffinches with papillomas. *Br. Birds* **58**, 346–347.

McDiarmid, A. (1962) Diseases of free living wild animals. *F.A.O. Agric. Stud.* No. 57. Rome.

McDiarmid, A. (1967). Some diseases of deer. *Deer* **1**, 78–89.

Owen, L. N. and Nielsen, S. W. (1968). In Press.

Shope, R. E. Mangold, R., MacNamara, L. G. and Dumbell, K. R. (1958). An infectious cutaneous fibroma of the Virginia white-tailed deer, *Odocoileus virginianus*. *J. exp. Med.* **108**, 797–802.

Snyder, R. L. and Ratcliffe, H. L. (1966). Primary lung cancers in birds and mammals of the Philadelphia Zoo. *Cancer Res.* **26**, 514–518.

Stewart, H. L. (1966). Comparison of histologic lung cancer types in captive wild mammals and birds and laboratory and domestic animals. *In* "Lung Tumours in Animals". (L. Severi, ed), pp. 25–58, Perugia: Division of Cancer Res.

Willis, R. A. (1960). "Pathology of Tumours". 3rd Edition. Butterworth's Press, London.

Woodford, M. (1966). Lymphosarcoma in a wild roe deer. *Vet. Rec.* **79**, 74.

Discussion

Cotchin (Chairman): This paper is now open for general discussion. Perhaps I might ask Dr. Jennings a question on the age of wild animals. It does seem to me that we need to know a lot more about age in respect of wild animals. Is anything known about the average age of dead wild animals? I understand that the age of red deer in Scotland can be determined fairly accurately and similar work, I believe, is going on with several other species.

Beer: Perhaps I might add a little about birds. On the question of water birds I don't think Dr. Jennings specifically mentioned the herring gull. These birds certainly live a long time; we have had quite a number between fifteen and twenty years old. There is a fair amount of data available from bird ringing on the ages to which birds survive. Recently we have heard quite a lot about increasing water pollution. I wonder if Dr. Jennings considers that perhaps there are significant amounts of carcinogens in the water these days possibly leading to tumours in water birds? This is pure speculation but a number of observers have noticed the susceptibility of waterfowl to tumours. In the larger zoos throughout the world incidents of respiratory cancer have increased over the last twenty years and it has been suggested this is due to an increased contamination of the water.

Taylor: Two of the previous speakers have mentioned the importance of examining African rodents for hepatic tumours. In 1962 I spent a short while in Kenya and during this period I examined several hundred multimammate rats which at that time were extremely numerous and were causing a lot of damage to agricultural crops. Unfortunately, I am no pathologist and although I looked inside these rats I was not particularly interested in disease. However, I did notice hepatic tumours in several animals and I would like to put this on record.

JENNINGS: This species is said to be very prone to tumours; these could well spread to the liver so investigations would have to be made as to whether these were primary or secondary tumours. It would be interesting if eventually the incidence of tumours in the wild species was found to be very much less than the corresponding incidence in captive animals. This would raise a number of points in one's mind as to what the cause of this difference could be.

Symp. zool. Soc. Lond. (1968) No. 24, 289–309.

SOME ASPECTS OF THE EPIZOOTIOLOGY OF MYXOMATOSIS

HELEN E. N. VAUGHAN and J. A. VAUGHAN

*Ministry of Agriculture, Fisheries and Food,
Infestation Control Laboratory, Field Research Station,
Worplesdon, Guildford, Surrey, England*

SYNOPSIS

In contrast to most wildlife diseases, myxomatosis is of importance because of the economic benefits it can produce. Following its deliberate introduction into Australia in 1950 and its arrival in Britain in 1953 after the release of the virus in France, myxomatosis has considerably reduced rabbit populations which were sufficiently high to be a major agricultural problem.

From the economic aspect, a study of the epizootiology of the disease is necessary to discover whether or not it will continue to provide an effective measure of control. The study of myxomatosis in the European rabbit is of particular interest because it also yields information on the evolving relationship between a pathogen and a new, highly susceptible host in two different and widely separated parts of the world. The two main features which have emerged are attenuation of the virus and the possibility of enhanced genetic resistance in the host.

In Australia, attenuation of the virus was observed within a year of the first epizootic and "strains" giving the original high mortalities are no longer recoverable from the field. The vectors, principally mosquitoes, have been shown to be a major influence on the selection and establishment in the field of these attenuated "strains". Genetic resistance also was soon detected and at one study area it has proceeded to a marked extent.

In Britain it would seem that attenuation, although marked, has not proceeded to quite the same degree as in Australia, possibly because a different vector, the rabbit flea, is involved. Genetic resistance does not yet appear to have increased considerably but studies at two areas indicate a lengthening survival time although there is no significant increase in the number of survivors. In Britain, at least, it may be concluded that myxomatosis is still having a significant influence on rabbit populations.

INTRODUCTION

In its natural leporid hosts, the South American *Sylvilagus brasiliensis* and the North American *S. bachmani*, myxoma virus appears to have reached a stable association and produces only a mild infection (Aragão, 1942; Marshall and Regnery, 1960). The outward symptom is a localized, benign tumour at the site of infection and this tumour can persist for many weeks providing a long-lasting source of infectivity for mosquito vectors.

Only when the virus is transferred to the European rabbit (*Oryctolagus cuniculus*) does it produce the severe, usually fatal, disease accompanied by the symptoms typical of "infectious myxomatosis" as originally described by Sanarelli (1898). A wide range of other verte-brates has been tested in the laboratory for susceptibility (Bull and Dickinson, 1937) and, apart from certain other North American *Sylvilagus* spp. (Fenner and Ratcliffe, 1965), no other species appear to be affected, with the exception of hares (*Lepus* spp.). A small number of instances have been reported of hares showing clinical symptoms of the disease (Jacotot, Vallée and Virat, 1954; Whitty, 1955; Collins, 1955; Anonymous, 1955). It is shown in this paper that specific antibodies can be detected in a high proportion of the members of certain hare popula-tions although no outward symptoms are apparent.

In contrast to most wildlife diseases, myxomatosis is of economic importance not because it is injurious to man's interests but because it has proved to be an important agent in the reduction of large rabbit populations, particularly in Australia, France and Britain. In these countries rabbits were present in sufficient numbers after the 2nd World War to form a major agricultural and forestry problem. In Australia the increased agricultural production during the first post-myxomatosis year alone was valued at £50 million. The benefit to agriculture and forestry in Britain was estimated to be not less than £45 million annually (Thompson and Worden, 1956).

From the economic standpoint, therefore, a study of the epizooti-ology of myxomatosis in the European rabbit is necessary to determine the extent to which it will continue to provide an effective degree of control. From the scientific aspect, the study of myxomatosis is of particular interest because it provides information on the evolving relationship between a pathogen and a new, highly susceptible host in two widely separated parts of the world, in each of which the attitude to the disease and its management has been very different. In Australia the disease was deliberately introduced in 1950 as a method of rabbit con-trol. In some states, particularly New South Wales, Western Australia and Victoria, a policy of active encouragement has been followed since then by promoting the repeated introduction of virulent strains of virus in annual inoculation programmes. By contrast, in Britain efforts were made to confine and eradicate the initial outbreaks which occurred in 1953 after the release of virus in France the previous year. When these measures failed, it was decided that the disease should be allowed to follow its natural course and delibrate re-introductions of the virus have not been made.

In both countries the two main features which have emerged, as the host and the pathogen become mutually adapted, are attenuation of the virus and the possibility of increasing genetic resistance. It is the con-sideration of these two features that forms the substance of this paper.

ATTENUATION

The strains of virus used to initiate the Australian and European epizootics were both of Brazilian origin and were of high virulence for *O. cuniculus*, producing 100% mortality in laboratory rabbits. Strains giving reduced kills were recovered from the field in Australia within a year of the initial release of virus (Fenner, 1953; Mykytowycz, 1953). In Europe, the first attenuated strains were reported from France and Britain in 1955, approximately two years after the first outbreak, (Jacotot *et al.*, 1955a; Fenner and Marshall, 1955; Hudson, Thompson and Mansi, 1955) although it was later shown that two strains recovered from an outbreak in Sussex late in 1954 were also of reduced virulence (Fenner and Marshall, 1957).

Fenner and Marshall (1957) demonstrated the existence of a negative correlation between the mortality rate and the mean survival time of a small group of rabbits for strains of differing virulence. This finding was used to develop a standard technique by which any virus strain could be assigned to one of five (later six) virulence grades, each covering a range of estimated mortalities. The method used was to inoculate intradermally six laboratory rabbits with a small standard dose of virus and then to note the symptomatology and measure the mean survival time. It has been pointed out by Marshall and Fenner (1960) that these virulence grades are merely a convenient way of handling data and have no genetical meaning.

Attenuated strains recovered from the field cannot be distinguished serologically from each other, or from their Brazilian originals, by means of the Ouchterlony gel diffusion test (Chapple, Bowen and Lewis, 1963; Fenner and Chapple, 1965) although Reisner, Sobey and Conolly (1963) were able to demonstrate differences between the soluble antigens of strains of Brazilian and Californian origin. A tissue culture neutralization test, described below, has also been used in an attempt to show serological differences between attenuated and virulent strains, but no significant differences have been noted (H. E. N. Vaughan and N. D. Lewis, unpublished). Therefore, it would appear that the method of Fenner and Marshall, using laboratory rabbits, is the only technique available at present for distinguishing between strains of differing virulence. Examination of virus recovered from the field in Australia revealed a progressive increase in the proportion of attenuated strains. By 1958–59 no fully virulent, Grade I, strains were recovered and the majority of strains fell into the Grade IIIA–IIIB group (Marshall and Fenner, 1960). By 1963–64 the trend towards attenuated strains had increased still further, with few Grade II strains being recovered and an increase in the proportion of Grade IV strains (Fenner and Ratcliffe, 1965) (see Table I). When fully-virulent virus was re-introduced in three study areas, it was found to be rapidly eliminated from the population in favour of strains of reduced virulence (Fenner, Poole, Marshall and Dyce, 1957). Fenner, Day and Woodroofe (1956) have shown that mosquitoes, which are the principal vectors in Australia, play a major part in the selection of attenuated strains. Mosquitoes readily transmit infection only when the virus titre in skin lesions has reached a high level and rabbits infected with moderately attenuated strains give the most favourable combination of virus-rich skin lesions and extended survival times. Rabbits infected with virulent strains soon die after the virus titre has reached a high level and, in rabbits infected with greatly attenuated strains, the virus titre may not reach very high values and falls away as the lesions regress. Thus, moderately attenuated strains such as Grades III and IV possess a selective advantage for transmission by mosquitoes and become dominant in any population.

In Britain it was soon shown that the principal vector was the rabbit flea, *Spilopsyllus cuniculi* (Dale) (Lockley, 1954; R. C. Muirhead-Thompson, unpublished), and that mosquitoes played little part except in certain specialized cases (Muirhead-Thompson, 1956a, 1956b). It was generally considered at this time that *S. cuniculi* was a sedentary type of flea, leaving the host only at the latter's death. This might be expected to favour selection of the more virulent strains since these, bringing about the death of the host in the shortest time, would cause the most frequent release of infective vectors. However, a large scale survey made in 1962, nine years after the initial outbreak, showed that the situation with respect to attenuation was very similar to that found in Australia at the comparable time, 1958–59 (Fenner and Chapple, 1965). The results of these surveys, together with the 1963–64 Australian survey, are shown in Table I.

It has since been shown (Mead-Briggs, 1964) that *S. cuniculi* is far more mobile than had been supposed and that a considerable interchange of fleas can occur between rabbits. This would tend to select against any preferential spread of the more virulent strains of virus and it is of interest to know whether a situation analogous to that found in Australia for mosquito vectors occurs for the flea vector in Britain. A laboratory study is at present in progress to determine the proportion of fleas able to transmit infectivity when taken from a myxomatous rabbit at intervals after its infection. Preliminary results, using a representative strain from each of the virulence Grades I, IIIA and V, indicate that the Grade IIIA strain is most readily transmitted, giving a high proportion of infective fleas for several days. If this similarity to the Australian

TABLE I

A comparison of the virulence of strains of myxoma virus recovered in Australia and Britain several years after its introduction (figures indicate percentages)

	Virulence Grade					
	I	II	IIIA	IIIB	IV	V
Estimated % mortality	>99	99–95	95–90	90–70	70–50	<50
Australia						
1958–59	0	24·6	29·2	26·1	14	6·1
1963–64	0	0	26·4	42·6	25·6	5·4
Britain						
1962	4·1	17·6	38·8	24·8	14·0	0·9

Date of the first outbreak in Australia—1950
Date of the first outbreak in Britain—1953
(after Fenner and Chapple, 1965)

findings can be confirmed, using more strains of these and other grades, then it would appear that, as in Australia, moderately attenuated strains possess a selective advantage and will remain dominant in the population.

No country-wide survey of virulence has been carried out in Britain since 1962 so it is not known whether the trend towards increasing attenuation shown in the Australian survey of 1963–64 (Table I) will be repeated here. However, the seventeen field strains from the mainland tested since 1962 have all fallen into the Grade IIIA. It may be that a higher titre of virus is required in the skin lesions for fleas to acquire infectivity than for mosquitoes and that, therefore, strains of a slightly lesser degree of attenuation will be selected.

ACTIVE IMMUNITY

The appearance of attenuated strains of virus gave rise to numbers of recovered rabbits and some populations have been examined for the presence of active immunity. Study areas were chosen in Norfolk, Kent, Berkshire and Yorkshire. Samples were taken throughout the year for periods of up to four years and were tested to determine seasonal fluctuations in proportions of immune, infected and susceptible animals.

Whole rabbit carcasses were obtained and where possible fresh blood samples. Portions of spleen, lung, any lesion material, blood and, when available, separated serum were tested by the Ouchterlony gel diffusion test for the presence of myxoma antigen or antibody. The separated antiserum was also tested by means of a sensitive tissue culture neutralization test carried out on monolayers of rabbit kidney cells, RK_{13}. Schwerdt and Schwerdt (1962) evolved a plaque method for the titration of the infectivity of myxoma virus on monolayers of rabbit kidney cells and showed that the plaque count was depressed when the virus was mixed with myxoma antiserum. A quantitative method of assaying neutralizing antibody using varying dilutions of virus and antiserum has been evolved (H. E. N. Vaughan and N. D. Lewis, unpublished). In comparative neutralization experiments, titration on monolayers of RK_{13} cells, on the chorio-allantoic membrane of eggs, and by the gel diffusion test showed the egg to be almost as sensitive as the RK_{13} monolayer but the egg to egg variation and non-specific reactions obtained made this method unacceptable for accurate and quantitative estimations of antibody titre. The tissue culture neutraliz-ation test proved to be the best method; it gave the most accurate quantitative estimates and was some 100 times more sensitive in the detection of neutralizing antibody than the gel test. The gel diffusion test failed to detect antibody in a significant proportion of samples from immune wild rabbits yielding positive results by neutralization test. Live rabbits, whose sera had low neutralization indices as determined by the neutralization test, showed a typical immune reaction on challenge; namely, the appearance on the second day after inoculation of a large primary lesion which regressed swiftly to form a scab. No other symp-toms were observed but an increase in antibody titre was noted. It was decided, therefore, to exclude all rabbits obtained from the study areas which were not accompanied by fresh blood samples. Thus only rabbits whose sera gave a positive reaction by the tissue culture neutralization test have been included in the immune category and only those whose sera gave a negative result have been classed as susceptible.

The results from the four study areas, given in Table II, show a similar pattern. In the winter and early spring, the populations were largely immune with the occasional infected rabbit and a small number of susceptible animals. In spring and summer, a greater proportion of susceptible rabbits was found and by the autumn the number of infected animals increased. By the beginning of the winter, the populations again consisted mainly of immune rabbits but with appreciable numbers of infected animals. Myxomatosis was typically present in each of the areas studied throughout the year, with peak incidence during an annual summer epizootic which is probably governed by the seasonal breeding of the rabbit. As the breeding season progresses, the number of young susceptible animals increases. Some become infected with virus which then spreads rapidly through this new and susceptible part of the

TABLE II

To show the seasonal fluctuation in the proportions of immune, infected and susceptible rabbits from four study areas (figures are total numbers of animals collected)

Seasons	Norfolk sample Totals for 1964–1967		
	Immune	Infected	Susceptible
Dec.–Feb.	99	2	2
March–May	44	10	71
June–Aug.	11	6	44
Sept.–Nov.	61	40	5
Seasons	Berkshire sample Totals for 1965–1966		
	Immune	Infected	Susceptible
Dec.–Feb.	19	2	2
March–May	1	1	1
June–Aug.	0	0	4
Sept.–Nov.	30	7	9
Seasons	Yorkshire sample Totals for 1965–1966		
	Immune	Infected	Susceptible
Dec.–Feb.	10	1	0
March–May	10	4	13
June–Aug.	0	1	0
Sept.–Nov.	82	28	2
Seasons	Kent sample Totals for 1964–1967		
	Immune	Infected	Susceptible
Dec.–Feb.	40	11	7
March–May	15	9	17
June–Aug.	16	20	51
Sept.–Nov.	29	21	21

population. By early winter, a largely immune population remains to form the breeding stock for the following year, together with a few susceptible rabbits which have escaped infection. These susceptible rabbits which escaped the summer epizootic provide the reservoir for the disease throughout the winter and by the following spring susceptible rabbits are rare.

Thus in Britain rabbit populations tend to be subjected to an annual epizootic, governed by the breeding cycle of the rabbit and not, as in Australia, by the breeding cycle and availability of vectors since in

Britain the rabbit flea is present on rabbits at all seasons. It is important to note that because the disease can persist throughout the year the chance of most rabbits becoming infected is maximal, a circumstance which favours the build-up of any genetic resistance.

GENETIC RESISTANCE

As a result of the occurrence in Britain and Australia of many strains of myxoma virus of markedly reduced virulence, a considerable number of survivors possessing active immunity to the disease has remained after each epizootic. Rabbits possessing any degree of genetic resistance will have a selective advantage and in both countries experiments have been made to try and detect the presence of any such genetic resistance.

Workers in Australia have tested the resistance of wild rabbits from study areas in New South Wales and Victoria at yearly or longer intervals (Marshall and Fenner, 1958; Marshall and Douglas, 1961; Fenner and Ratcliffe, 1965). Large numbers of 4–6 week-old rabbits were captured alive and kept in animal houses until they were at least four months of age. By this time they were thought to have lost any immunity passively acquired from the mother and to react to infection as adults (Fenner and Marshall, 1954). The sera of these animals were tested for the presence of antibody and actively immune rabbits were excluded. The remaining susceptible animals, together with groups of normal laboratory rabbits, then received an intradermal inoculation containing 5 RID50 of a standard batch of KM_{13} strain of virus (Marshall and Fenner, 1958). The disease symptoms, mortality rates and survival times of these rabbits were noted. The KM_{13} strain, the prototype of Grade III virulence giving a mortality of 95–90% and m.s.t. of 17–22 days in laboratory rabbits, was chosen as it was thought that the very high virulence of the standard laboratory strain would obscure minor changes in host resistance.

Enhanced genetic resistance was found in several study areas and at one site, Lake Urana, New South Wales where the response of the rabbit was studied over a period of seven years, the mortality rate fell from 90% to 25% and the proportion of cases showing severe disease symptoms fell from 93% to 54% (Marshall and Fenner, 1958). In other areas the degree of genetic resistance developed, although significant, was not so great. It was suggested that the Lake Urana results might be related to the fact that this area had experienced regular annual epizootics, whereas epizootics had occurred at less frequent intervals in the other areas.

In Britain tests for genetic resistance have been made on rabbits from two areas where populations were reported to be increasing, one near Downham Market in Norfolk and one near Edenbridge in Kent. The method of testing was the same as that used in Australia with some modifications. Susceptible adult rabbits from both areas were included in the sample. The sera of these rabbits was examined for the presence of antibody by the tissue culture neutralization test and as it has been found that the test is capable of detecting antibody in the sera of wild rabbits kept in captivity for three years (H. E. N. Vaughan, unpublished) the inclusion of adult susceptible rabbits in these studies is justified. Large numbers of very young rabbits were obtained from both areas and, after a period of isolation in the animal house to detect any animals incubating infection, they were released into outdoor pens for several months. The mortality of these juveniles was high during the first two years of the study but it was greatly reduced when the young rabbits were first kept in individual cages and supplied with plentiful fresh greens and allowed to become accustomed to a pellet diet (SG 41) before being released in small numbers into compounds provided with adequate greens, water and pellets. This difficulty in rearing young rabbits, also mentioned by Marshall and Fenner (1958), accounts for the small numbers of results in some years.

These animals were returned to the animal house when they were 4–5 months old and weighed approximately 1000 g. Two weeks later each received an intradermal inoculation of 50 RID50 of the Brecon strain of virus. This strain of virus, recovered from an infected rabbit from Breconshire (Chapple and Bowen, 1963), produces similar symptoms and survival times to the KM_{13} strain used in Australia. The dosage of 50 RID50 was chosen as it was found in earlier experiments that wild rabbits did not always react to the dose required to infect domestic rabbits (5–10 RID50) although normal symptoms followed the inoculation of a larger dose of the same virus. Batches of New Zealand White domestic rabbits and wild rabbits from the Island of Skokholm also received the same inoculation. The former served as the "domestic baseline" to ensure that the virulence of the virus sample remained constant and the latter as a "wild baseline" against which the "domestic baseline" and the reactions of the wild populations from Norfolk and Kent could be compared. Rabbits from Skokholm, an island off the Pembrokeshire coast, were chosen as a "wild baseline" since, apart from some unsuccessful attempts to introduce the virus (Lockley, 1940), myxomatosis was not recorded there until 1967. Such rabbits were considered to comprise a genetically unselected population against which samples from the study area could be compared. All inoculated rabbits were examined daily for the appearance of disease symptoms and survival times and numbers of survivors were recorded. The results are shown in Table III and Fig. 1.

TABLE III

Genetic resistance study. The mean survival times, range of survival time and number of survivors obtained for "domestic" and "wild" baselines and rabbits from the two study areas

Year		Domestic baseline	Wild (Skokholm) baseline	Kent sample	Norfolk sample
1964	No. tested	5	5		8
	m.s.t. (Days)	21·8	25·4	—	24·4
	Range (days)	17–27	19–29		19–30
	No. survivors	0	0		2
1965	No. tested	32	43	17	5
	m.s.t. (days)	17·3	23·1	21·7	23·3
	Range (days)	13–21	15–36	12–31	20–27
	No. survivors	0	1	2	0
1966	No. tested	23		26	41
	m.s.t. (days)	18·9	—	27·5	30·2
	Range (days)	14–27		16–35	19–49
	No. survivors	0		0	4
1967	No. tested	16	40		34
	m.s.t. (days)	19·2	23·0	—	23·7
	Range (days)	16–26	15–31		18–39
	No. survivors	0	1		2

A virulence test was made on samples from infected animals from both study areas during each year of the surveys. In all cases the virus was of Grade IIIA virulence.

There were marked differences between the "domestic" and "wild" baselines and Table III shows that the wild rabbits had a prolonged survival time and there were two survivors. As indicated in Table IV the time of appearance of symptoms was also different in the two sets of animals. A primary lesion was not always seen at the site of inoculation on the wild rabbit and sometimes the first sign of disease was swelling of the eyelids. All symptoms tended to appear later in wild than domestic rabbits and generalization was also less pronounced. Thus the New Zealand White "domestic baseline" can be used as an indicator for the stability of the test virus but not as a baseline for detection of genetic resistance. It is thought that the "wild Skokholm baseline" may be used as a basis for detection of genetic resistance in populations of wild rabbits exposed to myxomatosis but the most satisfactory method for detecting its development in a population is to obtain samples annually and to compare their mean survival times following infection by standardized procedure.

The results in Table III indicate that little genetic resistance has developed in the populations at either study area. Two of the 88 Skokholm rabbits survived, compared with 8/88 from Norfolk and 2/43 from Kent. Also the mean survival time for rabbits from both study areas are similar to those obtained for the Skokholm wild baseline. The results vary little from year to year, so far, but it will be interesting to compare these with results for subsequent years. Figure 1 shows an increase for both areas in the number of rabbits surviving for a longer time and also shows that in some samples many rabbits survive longer than those in the Skokholm population (e.g. Norfolk, 1966). The results also indicate the presence of a mixed population with respect to resistance in both study areas, some rabbits showing no resistance to disease and behaving like the baseline rabbits, some showing slight resistance and a small number having a greater degree of resistance as indicated by the markedly increased survival times. The overall picture is that which would be expected if genetic resistance is just entering a population.

In the two areas studied genetic resistance appears to be slight. Although genetic resistance may be beginning, as suggested by the increased survival time of many rabbits, the absence of any increase in the number of survivors in successive years indicates that it is not yet of significant importance in the build up of these rabbit populations. It is hoped that this work will be continued in these and other areas, as a yearly assessment of the susceptible population and a comparison of results is the only way in which the growth of genetic resistance can be determined.

An interesting question is thus posed. Why has no significant degree of genetic resistance arisen in these areas where the rabbit populations experience enzootic disease conditions with an annual summer epizootic? These are the circumstances which it is thought produced the rapid increase in the genetic resistance found at Lake Urana. It is known that the high temperatures experienced during the Australian summer have a marked "sparing" effect, reducing the number of mortalities (Marshall and Douglas, 1961) but it seems unlikely that this is the full explanation of the difference.

TABLE IV

The time of appearance of disease symptoms in three groups of New Zealand White domestic rabbits and Skokholm wild rabbits following infection with similar doses from the same batch of myxoma virus (Brecon)

| | Primary lesion | | Secondary lesions | | Eyes affected | Eyes closed | Range of survival times |
	Time of appearance (days)	No. showing symptom	Time of appearance (days)	No. showing symptom	Time of appearance (shown by all animals) (days)		(days)
Skokholm I	6–7	$\frac{5}{6}$	8–12	$\frac{5}{6}$	8–10	12–15	19–32
Domestic I	3	$\frac{6}{6}$	6–8	$\frac{6}{6}$	9–11	11–13	17–27
Skokholm II	5–6	$\frac{9}{18}$	7–8	$\frac{17}{18}$	9–10	13–17	17–31
Domestic II	3–4	$\frac{6}{6}$	4–5	$\frac{6}{6}$	7	10–11	16–25
Skokholm III	4–6	$\frac{8}{22}$	6–7	$\frac{20}{22}$	7–11	11–15	17–30
Domestic III	3–5	$\frac{5}{5}$	6	$\frac{5}{5}$	7	10	18–21

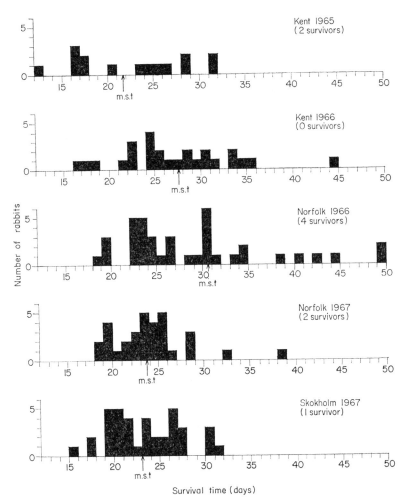

FIG. 1. A comparison of survival times of wild rabbits from the two genetic resistance study areas and Skokholm Island, following infection with a standard dose of the Brecon strain of myxoma virus. (m.s.t.: mean survival time.)

MYXOMATOSIS AND HARES

Asmentioned in the introduction there have been a few reports of hares showing clinical symptoms of myxomatosis and it has been suggested that hares might act as a reservoir of virus and perhaps transfer the disease from one rabbit population to another. However, no attempts have been made to examine large numbers of animals for the presence of antigen or antibody.

In 1964 samples of clotted blood from 172 hares were examined by the gel diffusion test for the presence of myxoma antigen or antibody (N. D. Lewis, unpublished). Five samples showed weak antibody, two weak antigen and the remainder were negative for both. In 1965 a further 95 hares from Cambridgeshire were examined. Tissues from the body and blood samples were examined by the gel diffusion test: antigen was detected in tissues from five hares and antibody in tissues from six. The separated serum obtained from the blood samples was also tested by the tissue culture neutralization test which showed that 30 hares had antibodies to myxoma virus. Thus 31% of the hares examined in this survey had been infected at some time with myxoma virus. None showed disease symptoms even in cases where antigen was detected. Following the results of this survey it was decided to try to obtain hares, inoculate them with large doses of virus and observe them closely to detect any disease symptoms.

Two unrelated hares (*Lepus europaeus occidentalis*) (one female and one male) were obtained in June 1964 when they were only a few weeks old. They were bled and the separated serum proved negative when tested for the presence of myxoma antibody by the neutralization test.

TABLE V

Serum neutralization indices obtained at various times following injection of two hares with myxoma virus

Days post inoculation		10	17	31	66	152	239
Serum neutralization indices obtained following first inoculation	Male Hare	1·2	1·8	1·8	2·1	1·0	0
	Female Hare	1·5	1·5	1·9	2·1	1·1	0

Days post inoculation		7	14	25	35	50	70	241
Serum neutralization indices obtained following second inoculation	Male Hare	0·9	1·4	1·5	—	1·5	1·7	1·0
	Female Hare	2·7	2·7	1·7	1·7	1·9	1·4	1·0

Nine months later both hares received an intradermal inoculation containing 10^5RID50 of the Cornwall strain of virus of Grade I virulence (Fenner and Marshall, 1957). The animals were examined daily for symptoms of disease and blood samples were taken at intervals to test for the presence of neutralizing antibodies. The two animals received a similar inoculation after one year and again blood samples were taken at intervals.

No primary reaction was produced at the site of inoculation nor was any other symptom of disease noted. As shown in Table V antibody was detected 10 days after the first inoculation and a maximum value was obtained shortly after this time. The antibody level then began to fall slowly and none could be detected after a 7 month period. After the second inoculation, the antibody level rose to a higher value and could still be detected 7 months after inoculation. The result of this single experiment and the surveys mentioned above confirm the work of Jacotot *et al.* (1955b) and suggest that although a substantial proportion of hares may become infected with myxoma virus, they very rarely show disease symptoms and it is unlikely that they play any part in transmission of the virus.

CONCLUSION

It is interesting to compare the present situation with regard to attenuation and genetic resistance in Australia and Britain. Despite the difference in major vector found in the two countries attenuation has proceeded along similar lines. Strains of virus of reduced virulence were first observed soon after the introduction of virus in both countries and within approximately nine years the moderately attenuated strains of Grade III virulence had become dominant. Since then attenuation has proceeded more slowly although in Australia an increase in the proportion of Grade IV strains has been recorded. In Britain, although another country-wide survey has not been made, there is some evidence to indicate that attenuation may not have proceeded to this extent. As mentioned above this may perhaps be related to the vector involved in transmission of the virus. In both countries the pathogen has become adapted to its new host in a comparatively short space of time and in both countries the vector has been shown to play a major part in this process of adaptation.

In contrast to the similarity in the progress of attenuation found in both countries, the situation with respect to genetic resistance appears to be very different. In Australia enhanced genetic resistance has been found in several study areas and at one area in particular a rapid rise in genetic resistance was shown over a period of seven years. In this country our studies have provided little evidence of enhanced genetic resistance. This is of particular interest since the populations studied experienced an annual epizootic caused by an attenuated strain of virus and the disease was present at a low level throughout the year. Both these factors, it has been suggested, would favour the build-up of genetic resistance by subjecting populations to a continuous selection pressure.

It is apparent that rabbit populations in this country are increasing only very slowly, despite an increase in the mean number of live young produced per doe in the years since the introduction of myxomatosis (Lloyd, 1963). The two main factors acting on the rabbit populations are predators and disease. The effect of the first is unknown, but can hardly account for the whole of this loss of young animals. It would appear then that myxomatosis is still playing a major part in keeping rabbit numbers at a low level and will continue to do so in the absence of genetic resistance.

ACKNOWLEDGEMENTS

We wish to thank Mr. H. V. Thompson and Dr. A. R. Mead-Briggs for their advice and help during the course of this work and for reading and criticizing the manuscript. We should also like to thank Messrs N. D. Lewis, J. Gurnell and K. D. West for technical assistance and Mr. H. Holman and several colleagues of the Ministry staff for obtaining material. We are grateful to Mr. M. Pratt and the late Col. E. Pratt for providing facilities for the collection of part of the material. The information in Table I is taken from the work of Fenner and Chapple originally published in the Journal of Hygiene in 1965 and is reproduced by kind permission of the Cambridge University Press.

REFERENCES

Anonymous (1955). Myxomatosis in hares. *Vet. Rec.* **67**, 455.
Aragão, H. de B. (1942). Sensibilidade do coelho do mato ao virus do mixoma; transmissão pelo *Aedes scapularis* e pelo *Stegomyia*. *Braz. méd.* **56**, 207–209.
Bull, L. B. and Dickinson, C. G. (1937). The specificity of the virus of rabbit myxomatosis. *J. Coun. scient. ind. Res. Aust.* **10**, 291–294.
Chapple. P. J. and Bowen, E. T. W. (1963). A note on two attenuated strains of myxoma virus isolated in Great Britain. *J. Hyg., Camb.* **61**, 161–168.
Chapple, P. J. ,Bowen, E. T. W. and Lewis, N. D. (1963). Some observations on the use of the Ouchterlony gel diffusion technique in the study of myxomatosis. *J. Hyg., Camb.* **61**, 373–383.
Collins, J. J. (1955). Myxomatosis in the common hare, *Lepus europaeus. Ir. vet. J.* **9**, 268–269.
Fenner, F. (1953). Changes in the mortality rate due to myxomatosis in the Australian wild rabbit. *Nature, Lond.* **172**, 228.

Fenner, F. and Chapple, P. J. (1965). Evolutionary changes in myxoma virus in Britain. An examination of 222 naturally occurring strains obtained from 80 counties during the period October–November 1962. *J. Hyg., Camb.* **63**, 175–185.

Fenner, F. and Marshall, I. D. (1954). Passive immunity in myxomatosis of the European rabbit (*Oryctolagus cuniculus*): the protection conferred on kittens born by immune does. *J. Hyg., Camb.* **52**, 321–336.

Fenner, F. and Marshall, I. D. (1955). Occurrence of attenuated strains of myxoma virus in Europe. *Nature, Lond.* **176**, 782–783.

Fenner, F. and Marshall, I. D. (1957). A comparison of the virulence for European rabbits (*Oryctolagus cuniculus*) of strains of myxoma virus recorded in the field in Australia, Europe and America. *J. Hyg., Camb.* **55**, 149–191.

Fenner, F. and Ratcliffe, F. N. (1965) "Myxomatosis". Cambridge University Press, London.

Fenner, F., Day, M. F. and Woodroofe, G. M. (1956). Epidemiological consequences of the mechanical transmission of myxomatosis by mosquitoes. *J. Hyg., Camb.* **54**, 284–303.

Fenner, F., Poole, W. E., Marshall, I. D. and Dyce, A. L. (1957). Studies in the epidemiology of infectious myxomatosis of rabbits. VI. The experimental introduction of the European strain of myxoma virus into Australian wild rabbit populations. *J. Hyg., Camb.* **55**, 192–206.

Hudson, J. R., Thompson, H. V. and Mansi, W. (1955). Myxoma virus in Britain. *Nature, Lond.* **176**, 783.

Jacotot, H., Vallée, A. and Virat, B. (1954). Sur un cas de myxomatose chez le liévre. *Annls Inst. Pasteur, Paris.* **86**, 105–107.

Jacotot, H., Vallée, A. and Virat, B. (1955a). Apparition en France d'un mutant naturellement atténué du virus de Sanarelli. *Annls Inst. Pasteur, Paris.* **89**, 361–364.

Jacotot, H., Vallée, A. and Virat, B. (1955b). Étude sur la transmission expérimentale de la myxomatose au liévre. *Annls Inst. Pasteur, Paris.* **88**, 1–10.

Lloyd, H. G. (1963). Intra-uterine mortality in the wild rabbit, *Oryctolagus cuniculus* (L) in populations of low density. *J. Anim. Ecol.* **32**, 549–563.

Lockley, R. M. (1940). Some experiments in rabbit control. *Nature, Lond.* **145**, 767–769.

Lockley, R. M. (1954). The European rabbit flea, *Spilopsyllus cuniculi* as a vector of myxomatosis in Britain. *Vet. Rec.* **66**, 434–435.

Marshall, I. D. and Douglas, G. W. (1961). Studies in the epidemiology of infectious myxomatosis of rabbits. VIII. Further observations on changes in the innate resistance of Australian wild rabbits exposed to myxomatosis. *J. Hyg., Camb.* **59**, 117–122.

Marshall, I. D. and Fenner, F. (1958). Studies in the epidemiology of infectious myxomatosis of rabbits. V. Changes in the innate resistance of Australian wild rabbits exposed to myxomatosis. *J. Hyg., Camb.* **56**, 288–302.

Marshall, I. D. and Fenner, F. (1960). Studies in the epidemiology of infectious myxomatosis of rabbits. VII. The virulence of strains of myxoma virus recovered from Australian wild rabbits between 1951 and 1959. *J. Hyg., Camb.* **58**, 485–488.

Marshall, I. D. and Regnery, D. C. (1960). Myxomatosis in a Californian brush rabbit (*Sylvilagus bachmani*). *Nature, Lond.* **188**, 73–74.

Mead-Briggs, A. R. (1964). Some experiments concerning the interchange of rabbit fleas, *Spilopsyllus cuniculi* (Dale) between living hosts. *J. Anim. Ecol.* **33**, 13–26.

Muirhead-Thompson, R. C. (1956a). The part played by woodland mosquitoes of the genus *Aedes* in the transmission of myxomatosis in England. *J. Hyg., Camb.* **54**, 461–471.

Muirhead-Thompson, R. C. (1956b). Field studies of the rôle of *Anopheles atroparvus* in the transmission of myxomatosis in England. *J. Hyg., Camb.* **54**, 472–477.

Mykytowycz, R. (1953). An attenuated strain of the myxomatosis virus recovered from the field. *Nature, Lond.* **172**, 448–449.

Reisner, A. H., Sobey, W. R. and Conolly, D. (1963). Differences among the soluble antigens of myxoma viruses originating in Brazil and Califoria. *Virology* **20**, 539–541.

Sanarelli, G. (1898). Das myxomatogene Virus. Beitrag zum Studium der Krankheitserreger ausserhalf des Sichtbaren (Vorläufige Mitteilung). *Zentbl. Bakt. ParasitKde* **23**, 865–875.

Schwerdt, P. R. and Schwerdt, C. E. (1962). A plaque assay for myxoma virus infectivity. *Proc. Soc. exp. Biol. Med.* **109**, 717–721.

Thompson, H. V. and Worden, A. N. (1956). "The Rabbit". Collins, London.

Whitty, B. T. (1955). Myxomatosis in the common hare—*Lepus europaeus*. *Ir. vet. J.* **9**, 267.

DISCUSSION

IRVIN: In Australia the mosquito is the principle vector of myxomatosis and in Britain, the flea. Is this because suitable fleas do not occur in Australia and suitable mosquitoes in Britain or is a different virus strain involved or is myxomatosis found in different ecological habitats in the two countries?

VAUGHAN: The rabbit flea does not occur in Australia and in this country mosquitoes, except in one specialized case, are restricted to certain areas. They have been found responsible for causing epizootics in coastal regions. I think you are right in saying that mosquitoes play a more important part in Australia because a suitable flea is not available.

CHAPMAN: Why mosquitoes do not play a more important part in this country I am not too sure. I would like to ask if you think there is any carrier state at all and do you think increased immunity could be due to a vector inoculating virus whilst the passive immunity derived from the mother is still present in the young rabbit. Alternatively, could the genetic resistance be due to the fact that they are inoculated each year in Australia and consequently more virus is available for the vector to inoculate into young rabbits before their passive immunity is lost?

VAUGHAN: I do not think there is any effective carrier state. The only source of virus for vector transmission appears to be the skin lesions and these are shed on recovery. You can tell a recently recovered rabbit by its partially bald nose and eyebrows but later on it is indistinguishable from an un-infected animal. I think passive immunity does play quite an important part in that animals infected as youngsters, say one month old, will recover; we have in fact had such animals, I believe from Berkshire, and as far as we could tell they could not have been more than six to eight weeks old. They had had the disease and completely recovered and they showed no outward signs of having been infected.

THOMPSON: Can you assess the virulence of myxomatosis virus and do you think there are enough susceptible rabbits to give an overwintering amount of myxomatosis. Do you think there are areas where they die out completely, say at the end of the autumn, and, finally, which are the strains isolated from the mainland?

VAUGHAN: All the strains from the mainland have been IIIA. I said from the mainland for a very good reason; until 1967, as far as we knew, myxomatosis was not on Skokholm but it is there now. How it got there we do not know and we don't know how the virus is maintained as the rabbit flea is absent from the island. Two samples have been brought back and tested for virulence; one was IIIA , the other IIIB. The Winter period is interesting as a small percentage of rabbits do breed throughout the year and this may provide sufficient susceptible animals to carry the infection over. Fleas are, I think, present on rabbits at all seasons of the year; you can always find them even although only in low numbers. I cannot help suspecting that there may be another important vector other than the flea.

ROTHSCHILD: As you know I have kept experimental rabbits for a number of years and I have had four outbreaks of myxomatosis among my rabbits very early in the year. On the first occasion it was in March and I was advised that it was too early for mosquitoes at that time. Later, in April I took some of my rabbits to the Royal Society which seemed to affect them very adversely because they got myxomatosis on their return. These rabbits are in cages two feet off the ground and I am very careful about introducing fleas from the outside world onto these rabbits; I am still mystified as to how these four outbreaks occurred so very early in the year particularly as the rabbits were kept in my own garden. Can anybody throw some light on this?

VAUGHAN: Any insect that will bite a rabbit and then bite a second rabbit can, I suppose, transmit myxomatosis. We have observed *Stomoxys* coming to our rabbit baits in large numbers but I think the spring is the wrong time of year for this particular species. I am not sure what particular flying vector one can get at that time of the year.

ROTHSCHILD: That is the mysterious part; fly vectors should not be about so early in the year and one does not expect to find such widespread myxomatosis amongst the wild rabbits at that time.

VAUGHAN: We are working on this subject at the moment and rabbits are being collected for us each week from one study area. Of course the main outbreak occurs in the autumn but one small area missed it and an epizootic occurred there this spring about the time when the flea population was rising on the rabbits; that might supply your reservoir of infection but again it does not tell us what the vector is. Are these rabbits from the outside; that is, wild ones?

ROTHSCHILD: No, they are tame rabbits and caged up; the fleas we use in our experiments are not brought in from outside so we have a mysterious vector somewhere.

VAUGHAN: I do not know much about mosquitoes in this part of the world but surely there are overwintering ones that come out of hibernation in the Spring and must, presumably, be biting.

CAPEL-EDWARDS: I was rather interested in your introductory remarks regarding the hare. I was not quite clear if these were purely clinical observations in the field or if any experimental work had been done on these animals. Could you please extend this point a little?

VAUGHAN: Quite a number of reports have appeared in the literature concerning myxomatosis in hares; material has been taken from lesions on hares and inoculated into rabbits which, in turn, have developed the disease. The occasional hare is susceptible to clinical myxomatosis. In one survey area ninety-three blood samples from hares were tested for the presence of neutralizing antibodies—thirty were positive although none of these hares showed any signs of clinical disease. We obtained two leverets about three weeks of age, reared them in captivity and inoculated them with massive doses of virulent virus and despite this they showed no external signs of myxomatosis. High antibody titres were, however, detectable; these fell away after about 250 days. The hares were then re-inoculated, the antibody titres rose again and again fell after the same length of time.

CHAIRMAN'S SUMMING-UP

We should thank Dr. Vaughan and Mr. Vaughan for their excellent presentation. This brings the session to a close. I don't think we ought to go away without letting Dr. McDiarmid know that we are very appreciative of his efforts in organizing and arranging this fine symposium, and we look forward to seeing the publication in due course.

AUTHOR INDEX

Numbers in italics refer to pages in the References at the end of each article.

SUBJECT INDEX

A

Adenoma:
  of kidney, 284
  pituitary, 283
Adiaspiromycosis, 257
Aedes africanus, 20
Aegyptianella pullorum, 189
Aepyceros melampus:
  foot-and-mouth disease in, 5, 6, 7
  trypanosome infection in, 149
Aeromonas
  liquefaciens, 40, 42
  salmonicida, 42
African horsesickness, 14
Agelaius phoeniceus phoeniceus,
    Plasmodium in, 180
Alces alces
  alces, 99, 230, 245
  americana, 230, 231
Allergy, 263
Alphitobius diaperinus (Darkling
    beetle), 281
Anas platyrhynchus (Mallard), 259
Antelope, saiga, *see Saiga tatarica*
Antilocapra americana, 232, 236
Apodemus
  agrarius, 113
  flavicollis, 82
  sylvaticus, 19, 76, 68, 82, 90, 121, 199,
    200, 202, 208
  Leptospira in, 82, 89, 92, 93
  pseudotuberculosis in, 113
  ticks on, 201
Apteragia quadrispiculata, 236
Arachnida, 199
Arbovirus, 13–28, 49
  arthropod hosts of, 14, 23
  cycle, 16
  defined, 13, 14
  distribution of, 23
  factors influencing survival of, 14
  factors influencing transmission of,
    15
  host-virus relationship, 14
  incidental hosts of, 14

Arbovirus—continued
  maintenance hosts of, 14, 18, 19, 22
  overwintering mechanism, 21–2
  threshold values, 15
  vertebrate hosts of, 14, 15, 16, 22,
    24
Argali, *see Ovis ammon*
Argentinian haemorrhagic fever, 21
Arthropods, haematophagous, 13
Artibeus, 21
Arvicola
  amphibius, 77, 199
  terrestris, 113
  leptospires in, 82
  pseudotuberculosis in, 113
Aspergillosis, 249, 250, 259, 268, 270
  in birds, 259
  pulmonary, in wild mammals, 260
Aspergillus, 268
  flavus, 259
  fumigatus, 259, 263
  nidulans, 259
  aflatoxins from, 264
Attenuation of myxomatosis virus,
    291, 303
Aythya marila, 249, 257

B

Babesia, 177, 178, 188, 189
  bigemina, 160
  divergens, 159, 160
  felis, 168, 169
  microti, 160, 170, 172
Badger, *see Meles meles*
Bats, 260, 262
  as virus hosts, 20–1
Beaver, salmonella in, 64
Behaviouristic studies, 51
Bicaulus alces, 233
Biochemical reactions of salmonella, 54
Birds:
  as virus hosts, 16–18, 22, 28
  aspergillosis in, 259
  blood parasites of, 179
  causes of death of wild, 276

321

DUE

(